Enablers for Smart Cities

Series Editor
Jean-Charles Pomerol

Enablers for Smart Cities

Edited by

Amal El Fallah Seghrouchni
Fuyuki Ishikawa
Laurent Hérault
Hideyuki Tokuda

iSTE

WILEY

First published 2016 in Great Britain and the United States by ISTE Ltd and John Wiley & Sons, Inc.

ISTE Ltd
27-37 St George's Road
London SW19 4EU
UK

www.iste.co.uk

John Wiley & Sons, Inc.
111 River Street
Hoboken, NJ 07030
USA

www.wiley.com

Library of Congress Control Number: 2016939641

British Library Cataloguing-in-Publication Data
A CIP record for this book is available from the British Library
ISBN 978-1-84821-958-8

Contents

Preface

Introduction

The concept of *smart cities* emerged few years ago as a new vision for urban development that aims to integrate multiple information and communication technology (ICT) solutions in a secure fashion to manage a city's assets. Modern ICT infrastructure and e-services should fuel sustainable growth and quality of life, enabled by a wise and participative management of natural resources to be ensured by citizens and government. The need to build *smart cities* became a requirement that relies on urban development that should take charge of the new infrastructures for smart cities (broadband infrastructures, wireless sensor networks, Internet-based networked applications, open data and open platforms) and provide various smart services and enablers in various domains including healthcare, energy, education, environmental management, transportation, mobility and public safety.

The smart enablers raise new research challenges that emerge across areas such as urban development and spatial planning, network infrastructure, technology platforms, services and applications, user behavior, cognitive modeling, service engineering, innovation theory and urban economics.

This book aims to present the most significant emerging, or already mature, research results in the domains mentioned below. This may help academics and practitioners to explore new directions and generate knowledge and solutions toward smarter cities.

This book has been written by experts and outstanding researchers in the main domains involved in smart cities' development.

Preface written by Amal EL FALLAH SEGHROUCHNI, Fuyuki ISHIKAWA and Kenji TEI.

The co-authors cover complementary domains of computer science needed such as cloud and distributed computing, artificial intelligence and sensors and belong to various and prestigious institutions in France, Japan, Italy and Spain, which bring rich experience and a broad overview of the topic.

This book consists of 11 chapters, covering a wide range of topics for smart cities. The chapters are ordered to go from technical foundations to enable certain advanced capabilities, to platform/middleware for supporting construction and execution of various applications to field experiences in cities. Each of these three parts and the involved chapters are self-contained and address issues in smart cities from various points of view. Thus, the reader can take any of the chapters independently according to his/her interest.

Chapters 1–4 present technical foundations to enable certain functions in smart cities. Specifically, the two capabilities of sensing and reasoning are focused on as the keys for smart cities. Sensing various kinds of context information is the starting point for smart support of human and societal activities. Then reasoning mechanisms make proper decisions about execution or deployment plans. Both of them need to consider the existence of heterogeneity regarding devices, information sources and applications. Chapters 1–4 address difficulties with this point as follows.

– Chapter 1 by Tei K. addresses how to manage multimodal sensor devices in wireless sensor networks collectively to support multiple applications with runtime task allocation and self-adaptation;

– Chapter 2 by Nakazawa J. addresses how various information sources on the Web, currently under limited use with Web browsers, can be transformed to sensor application programming interfaces (APIs) or data streams ready for easy integration into various application programs;

– Chapter 3 by Chaouche A. *et al.* addresses how smart behavior can be realized through spatiotemporal reasoning, not only by contextual planning but also by learning from past experiences;

– Chapter 4 by Piette F. *et al.* addresses how to dynamically use distributed smart applications over devices with heterogeneous resources with a special respect about privacy concerns.

Chapters 5–9 present platforms/middleware to support construction and execution of various applications. On one hand, this concerns how to support design, programing and verification of smart city systems. On the other hand, this concerns the architectural design in order to be holistic or to support specific features for smart city systems. Chapters 5–9 provide a wide range of discussions over these points.

– Chapter 5 by Tei K. addresses how to define the holistic architecture that supports a variety of functions necessary in smart city systems, such as virtualization, resource management and service composition;

– Chapter 6 by Gürgen L. *et al.* addresses how to support construction of event-driven behavior of smart city systems, which also requires proper configuration with devices;

– Chapter 7 by Ishikawa F. *et al.* addresses how to support verification of event-driven behavior of smart city systems, which can easily lead to conflicts between multiple devices, users and applications;

– Chapter 8 by Galache J. *et al.* discusses how to realize integration of heterogeneous devices for various smart city applications, with their experience in the Santander city, Spain;

– Chapter 9 by Olaru A. *et al.* discusses how to design the middleware tailored for smartness or intelligence provided on user devices at the right time.

Chapters 10–11 present field trials in cities. It focuses on making use of technology to realize values in cities and citizens' lives.

– Chapter 10 by Sotero Muñiz S. *et al.* reports their experiences in Santander, Spain, including city management by means of participatory sensing and traffic mobility management;

– Chapter 11 by Yonezawa T. reports the author's experiences in Fujisawa, Japan, including air-pollution monitoring using garbage trucks, delivery of live city information and tourist guidance by "smile coupon".

Although each chapter can be read independently, it is worth being aware of the connections between chapters. There is a specific emphasis on two aspects: practical integration and intelligence, both over heterogeneous devices, services and applications. Regarding the first aspect, some of the chapters come from the ClouT project, which aims at providing the infrastructure/platform of Cloud and Internet of Things for smart cities, as well as field trials. Chapters 1, 2, 5, 6, 7, 8, 10 and 11 are from the project (the whole picture of the project is found in Chapter 5). These chapters focus on how to integrate various devices or services to compose smart-city applications or experiences in cities on the basis of the integration. Regarding the second aspect, Chapters 3, 4 and 9 complement the other chapters by providing discussions on advanced techniques for smartness by multi-agent systems. This aspect is essential for emerging smart-city systems that embed interesting "human-like" characteristics of autonomy, sociality, adaptability and so on.

Biographies of the editors

Amal El Fallah Seghrouchni

Amal El Fallah Seghrouchni is a Full Professor at the University of Pierre and Marie Curie (Paris 6 – France) where she heads the MAS team at LIP6 (Laboratory of

Informatics of Paris 6). Her research is focused on autonomous agents and MAS. She is interested in the design, analysis and verification of cognitive MAS.

Her research topics find a large number of applications in the fields of design of complex systems (e.g. coordination of UAVs, multi-agent planning of aircraft missions etc.), ambient intelligence (smart-room, smart campus projects) and smart cities. These topics and applications are supported by a large number of academic collaborations and industrial projects. She has been involved in the major conferences related to autonomous agents and multi-agent systems and has published more than 150 referred papers and co-edited 16 books or post-proceedings.

Fuyuki Ishikawa

Fuyuki Ishikawa is an Associate Professor at National Institute of Informatics, Japan. He has worked on smart and trustworthy software engineering with emphasis on service-oriented computing and formal methods. His research experience includes leadership in several funded projects and publication of more than 100 referred papers about service composition, formal refinement, test generation, legal compliance and industry education.

Laurent Hérault

Dr. Laurent Hérault received his BS degree in electrical engineering and his MS degree in control engineering from the Grenoble Institute of Technology in 1987, followed by a PhD in computer science in 1991. He won the Best Junior Researcher Award from the University of Grenoble, France, in 1990. He is a CEA fellow since 2014, and led the Wireless Communications and Security labs from 2009 to 2011. Since 2011 he is VP, Director of the Europe division at CEA-LETI.

Hideyuki Tokuda

Hideyuki Tokuda obtained his PhD in computer science in 1983 from the University of Waterloo, Canada. He is a Professor of the Graduate School of Media and Governance and a Professor at the Faculty of Environment and Information Studies, Keio University, Japan. His current field of research is Information Appliance and Smart Spaces for the Ubiquitous Computing Environment. His research lab is working on applications, middleware, networks, and hardware to realize "Smart Spaces".

Acknowledgments

The idea of this book emerged from a collaboration of more than 10 years between the Honiden-Lab at the National Institute of Informatics (NII), Japan, and the SMA team (Systèmes Multi-Agents) at LIP6 at Université Pierre et Marie Curie (Paris 6), France. The editors would like to thank all the participants of the collaboration, especially those who contributed as the authors and the reviewers of the chapters.

We thank the consortium of the ClouT (Cloud of Things for empowering the citizen clout in smart cities) project that provided interesting insights and experiences in this book. The ClouT project is supported by the seventh framework programme for research (FP7) in the European Union (EU) and National Institute of Information and Communications Technology (NICT) in Japan.

Last but not least, we are very thankful for ISTE Ltd. and Professor Jean-Charles Pomerol who gave us the opportunity and support for publication.

Introduction

Chapter 1: *Shared wireless sensor networks as an enabler for a context management system in smart cities*. This chapter by Tei K. proposes middleware for Wireless Sensor Networks (WSNs), which provides the essential infrastructure for smart cities to have sensing capabilities, flexibly used anywhere. The focus of the middleware is shared by the use of WSNs: each sensor has several capabilities of sensing, and the sensors can be collectively used to allow for multiple applications to run. The middleware addresses difficulties because of the distributed and dynamic nature by means of a multi-level task description language as well as mechanisms for runtime task management and self-adaptation.

Chapter 2: *Sensorizer: an architecture for regenerating cyber physical data streams from the Web*. This chapter by Nakazawa J. proposes architectures for enhancing the sensing capability of smart city systems, by excavating information sources on the Web. Currently, there are various kinds of information on the Web only consumed by Web browsing of human users. The proposed approach enables what is called the "Sensorizer", which means making the information on the Web into sensor data streams easily accessible by application programs. The chapter shows how it is easy to sensorize existing Web information by a browser extension and also reports their experience with parking lot occupancy.

Chapter 3: *Smart agent foundations: from planning to spatio-temporal guidance*. The chapter by Chaouche A. *et al.* proposes an approach based on intelligent agents to build applications for smart cities. It highlights how software agents can be designed to assist users in their tasks and objectives in the context of the smart-city. These agents can be viewed as an interface between the user/citizen and the computational world of the smart city. The proposed approach is original and relevant to smart application design, as the agents can react on the fly, to the changes

Introduction written by Amal EL FALLAH SEGHROUCHNI, Fuyuki ISHIKAWA and Kenji TEI.

of contexts, sometimes unexpected, of their environment. A scenario of an ambient intelligent system dedicated to a smart campus is presented as well as the whole approach including BDI model of the agent, the planning process and the learning mechanisms that help to improve the agent performances and hence the quality of the smart applications.

Chapter 4: *A multi-agent middleware for deployment of ambient applications*. This chapter by Piette F. *et al.* proposes a middleware to ease the development, deployment, configuration and monitoring of applications for ambient systems. The main advantage of this middleware is that it decorrelates applications from hardware infrastructures by separately describing with different levels of the various entities of the system. The specifications and properties of the available hardware entities of the infrastructure, and the requirements of ambient applications are modeled with graphs. These descriptions allow us to reason about the deployment of applications on a heterogeneous hardware infrastructure by using a graph matching algorithm that finds a graph homomorphism between the application graph and the hardware infrastructure graph. The approach is based on multi-agent systems paradigm (MAS). The agent organization ensures data and resources privacy. Infrastructure agents are geographically located and manage a part of the hardware infrastructure. Application agents manage the applications used on it and the data generated by these applications.

Chapter 5: *ClouT: cloud of things for empowering citizen's clout in smart cities*. This chapter by Tei K. overviews the ClouT project, which investigates the power of combining Cloud and IoT (Internet of Things) for smart cities, through Europe–Japan collaboration. The project involves a conceptual definition to handle various kinds of things in a unified way following the Cloud model. The project also proposes a holistic architecture that involves various functionalities for the infrastructure and platform for smart cities. Notable techniques and field trials with cities are found in other chapters of this book (Chapters 1, 2, 6, 7, 8, 10 and 11).

Chapter 6: *sensiNact, IoT Platform as a Service*. This chapter by Gürgen L. *et al.* proposes the sensiNact framework to support the construction of IoT applications. The framework includes a platform to enable easy access to various kinds of devices. The key here is to handle the heterogeneity: various protocols with the application side and with the device side. The framework also includes a studio or integrated development environment (ID) to support description of event-driven behavior to sense and act using the devices. Thus, the chapter gives a good overview of how to support application construction with various IoT devices.

Chapter 7: *Verification and configuration of smart space applications*. This chapter by Ishikawa F. *et al.* proposes a framework for verifying and configuring smart-space applications. As depicted in Chapter 6, the behavior of smart applications is often event driven, often specified in terms of event–condition–action rules. These rules can easily cause conflict between one another but they are difficult to detect as they are hidden in a great many possible scenarios. The proposed framework facilitates using the model checking technique, which exhaustively checks possible transitions, by providing translation between engineer-familiar descriptions and mathematical descriptions for model checking.

Chapter 8: *SmartSantander, a massive self-managed, scalable and interconnected IoT deployment*. This chapter by Galache J. *et al.* reports their experience in Santander, Spain, regarding the architecture that supports deployment of various applications and services using IoT devices. The experience, supported by two projects of SmartSantander and ClouT, includes a unique massive deployment of IoT devices and field applications involving transportation monitoring, participatory sensing, augmented reality and so on.

Chapter 9: *Using context-aware multi-agent systems for robust smart city infrastructure*. This chapter by Olaru A. *et al.* focuses on how to achieve the goal of providing smart city users with fresh, relevant information, promptly, without the users needing to offer personal information in exchange, except when the information is absolutely necessarily and the user is fully aware of the transfer. The chapter proposes MAS-based architecture for context-aware AmI applications in which context is a first-class entity in the design of the system. The MAS uses a fully distributed context management architecture that requires no centralized components and relies as much as possible on the computational resources that belong to the user (rather than to other organizations). The DESIA architecture integrates the user's social context and the tATAmI framework/S-CLAIM language, which are the foundation for a middleware for the management of the user's data in a context-aware manner.

Chapter 10: *Santander City*. This chapter by Sotero S. *et al.* reports their experience in Santander, Spain, the same city as Chapter 10, but from the city government's side. This chapter, therefore, gives interesting insights about requirements and applications that should be supported by enabler techniques such as those described in the previous chapters. Specifically, the chapter reports smart-city applications for city management and traffic mobility management by using not only physical sensors, but also participatory sensing.

Chapter 11: *Fujisawa, towards a sustainable smart city*. This chapter by Yonezawa T. reports their experience in the Fujisawa city, Japan. The city investigates smart city applications, but in a sustainable way, that is, not accepting "introducing a lot of devices!" Their interesting idea is to attach sensors to garbage trucks to realize the sensing capability that covers the whole city. This chapter also reports other ideas to encourage, involve and interact with people by means of dashboards in stations and a mechanism with a "smile" game.

1

Shared Wireless Sensor Networks as Enablers for a Context Management System in Smart Cities

Wireless sensor networks (WSNs) are commonly used as a sensing infrastructure for smart city applications. A WSN is easy to use and can cover a wide area at low costs because of its wireless communication capability. The sensor nodes constituting a WSN are usually equipped with one or more sensor devices and can be used for different measurement purposes by reprogramming them. If WSNs could be shared by different smart city applications, they could be even more valuable enablers for smart cities. However, it is not easy to share WSNs. A shared WSN needs to support different kinds of measurement tasks at the same time and be able to accept new tasks at runtime. Even in a traditional closed WSN, its software should be carefully developed to satisfy certain quality requirements despite the severe resource constraints affecting the individual programmable sensor nodes (the sensor nodes of WSNs usually have quite limited resources, e.g. small batteries, low-spec CPU and narrow bandwidth). This issue is much harder to resolve in the case of a shared WSN. To satisfy the quality requirements of different applications, a WSN should be configured carefully according to specifications of the tasks, their quality requirements, and the environment, and should adapt its configuration in response to changes in the environment and the applications. A shared WSN should support various measurements, manage tasks at runtime and adapt to changes in the environment to reduce unnecessary consumption of resources. To develop such a shared WSN, we propose a middleware support for the network. In this chapter, we describe the architecture of our XAC middleware and the issues relevant to the shared WSN from the viewpoints of the task-description language, runtime task management and self-adaptation.

1.1. Introduction

In the smart cities of the future, many context-aware applications will support the citizen's activities by proactively controlling the various devices used therein.

Chapter written by Kenji TEI.

Context-aware applications will recognize the current context of the city they are monitoring and actuate devices to amend their status. A key service in smart cities will be context management systems, which estimate context of cities and provide it to applications.

A context management system should be able to collect and update various types of content required by context-aware applications and should be able to be used easily in various environments. Here, a wireless sensor network (WSN) will be a key infrastructure in context management systems. A WSN is a wireless *ad hoc* network consisting of tiny computers equipped with sensors and wireless communication devices. It continuously records and produces data by measuring the environment via sensor nodes. It can produce one or more kinds of data, because its nodes are equipped with one or more kinds of sensors, and it can be programed to alter or switch tasks between the different sensor devices used for monitoring. Moreover, it can be easily used because it does not require any communication cables. Its nodes communicate with each other via wireless links and transmit the measured sensor data via multi-hop communications.

These features of the WSN are quite important for context management systems in smart cities. First, its "easy-to-deploy" feature is suitable for the smart cities. Sensor nodes are usually used in outdoor spaces, but it is not easy to connect sensor nodes using cables because of monetary and legal constraints. Second, the "reprogrammable" feature is suitable because a smart city usually hosts many applications that require different kinds of sensor data. A context management system should carefully balance the demands of these applications and the resource consumption of the sensor nodes. This can be realized by reprogramming a WSN. Therefore, the WSN is a key enabler for a context management system in a smart city.

A shared WSN for smart cities should:

1) support various kinds of measurements;

2) manage tasks at runtime;

3) adapt to changes in the environment to reduce unnecessary resource consumption.

A shared WSN is used by many context-aware applications, which require different kinds of sensor data and different levels of accuracy. Therefore, it should be able to handle various measurements to produce one or more kinds of sensor data required by these applications with a level of accuracy. Moreover, applications using a WSN appear and disappear at runtime. Therefore, the WSN should be able to add or remove tasks at runtime without having to stop and start. Finally, a WSN should be able to adapt its behavior in response to changes in the environment. A WSN has severe resource limitations because each node in a WSN has CPU, memory, bandwidth and battery restrictions, and resources must be saved to increase the number of tasks that it can handle and to prolong its lifetime. Therefore, the WSN needs to automatically adapt to changes in the environment to reduce unnecessary resource consumption, that is to say without human intervention.

To develop such a shared WSN, middleware supports are needed. This chapter describes an example of middleware for a shared WSN, called XAC middleware. In addition, we discuss the research issues related to shared WSNs and the techniques used in XAC middleware.

1.2. Background

WSN software development is not easy because it requires programmers to have an in-depth knowledge of various fields, such as analysis of sensor data, distributed programing in wireless *ad hoc* networks and optimization of embedded systems. This section presents examples of types of WSN software to identify the issues concerning shared WSNs.

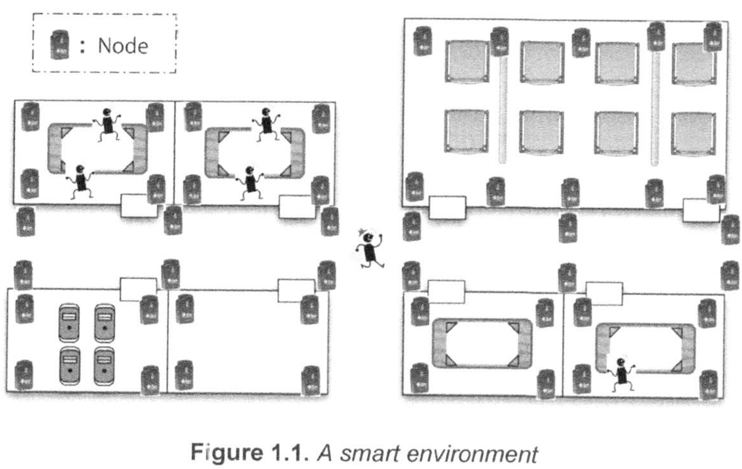

Figure 1.1. *A smart environment*

Let us first define our example– a smart environment is set up in an office building, where context-aware applications are introduced to optimize everyday business tasks. Consider the environment illustrated in Figure 1.1. Sensor nodes are used throughout rooms, corridors and stairwells to enable monitoring and to establish the current context of the building. Data sensed by the nodes are transferred via multi-hop communication to a central server (called the *base station* from here on).

Table 1.1 shows four scenarios, S1, S2, S3 and S4, envisioning context-aware applications in this example environment. In scenario S1, an application maintains the temperature levels in the conference rooms according to the preferences of the people in the room. The application in S2 determines the occupancy of conference rooms on the basis of the presence of people in the room and the reservation data of the room. Scenario S3 involves tracking applications that continuously monitor the current locations of staff inside the building, whereas the application in S4 detects suspicious intruders.

Scenarios	Application name	Operational tasks	Environmental information	Accuracy requirement
S1	Temperature management	Adjust room temperature according to preferences of people in the room	Temperature in the room	Within 2°C of actual value
S2	Meeting-room management	Maintain occupancy of conference rooms based on their current occupancy and reservations	Presence of people in conference rooms	Determine correct room occupancy with 99% accuracy
S3	Staff-tracking management	Determines the current locations of staffs	Location of staff	Within 1 m range in a public space or room
S4	Intruder detection	Determines suspicious intruders for instance by raising an alarm if people remain near an access lock for long periods without authenticating	Presence and position of people in certain locations	Within 2 m range

Table 1.1. *Examples of context-aware application scenarios*

Each application requires environmental context information related to its own operational tasks. S1 requires temperature data, S2 requires data on the presence of staff in each room, S3 requires data on the location of each member of the staff and S4 requires the location information of people in designated areas.

As we can see, each scenario possesses different non-functional requirements in terms of accuracy. Generally speaking, sensor data include a certain level of sensor error. A well-known way to improve accuracy is to aggregate sensor data coming from neighboring sensor nodes. For instance, knowing only the rough locations of staff (like the room in which a person is currently located) is enough for S3. In this case, the low accuracy requirements can probably be satisfied with sensor data from just one or two nodes. On the other hand, S4 requires the specific positions of staff to accurately track them within 2 m. This in turn entails gathering more sensor data than in S3.

Although many sensors are required from the viewpoint of accuracy, resource usage in a WSN should be kept as low as possible to prolong the lifetime of the network. Software developers should take into account the severe resource limitations of nodes in terms of CPU power, memory, communication bandwidth and so on when creating a WSN. In particular, the battery is a precious resource. For example, on average the battery of the commonly used Crossbow Mica2 node will deplete in just 7 days by reading its temperature sensor value and sending transmissions every second [SHN 04]. Even though the lifetimes of sensor nodes are gradually increasing as a result of hardware improvements, energy consumption is still an important issue in WSNs. Load concentrations on specific nodes will drain batteries quickly, which are then hard to recharge during runtime.

As such, to extend the network lifetime, it becomes necessary to use various optimization methods to extend each node's operational time, for instance, by aggregating sensor readings before transmitting them, by adjusting the sensing frequency to meet certain accuracy requirements or by duty-cycling node operation. To yield optimal results, these methods also need to comply with the requirements of multiple applications.

1.3. XAC middleware

XAC middleware is a middleware for a shared WSN. Its main features are as follows:

– WSN as a multi-modal sensor: XAC middleware uses the fact that a WSN is a multi-modal sensor. It hides the low-level details of the WSN from context-aware applications. A WSN can therefore be seen as a single sensor covering a large area by these applications. XAC middleware also provides a way to use the WSN for different measurement purposes. One or more applications can use the WSN at the same time.

– Runtime management: XAC middleware allows context-aware applications to register or unregister their measurement tasks during runtime.

– Self-adaptation: XAC middleware monitors changes in the WSN and adapts configurations in response to these to reduce unnecessary resource consumption and to maintain the required level of accuracy.

1.3.1. *Architecture of XAC middleware*

XAC middleware is implemented using SunSPOT, a commercial sensor node provided by Oracle. A WSN application with SunSPOT consists of base-station-side software running on a Java VM, and sensor-side software running on a Squawk VM in each node. Thus, the middleware consists of base-station-side middleware and sensor-side middleware.

The base-station-side middleware runs on the base station. It provides interfaces to context-aware applications to register or unregister their measurement tasks. In response to requests from the context-aware application, it either inputs software components relevant to a measurement task into the sensor side or removes them. The inputted components are used on sensor nodes designated by the deployment policy, and they report sensor data to base-station-side middleware. The base-station-side middleware provides the sensor data to the context-aware applications or initiates events by analyzing the sensor data in accordance with the specifications defined by each application. It also monitors the sensors and changes the configuration if it decides it is necessary to do so.

The sensor-side middleware runs on each sensor node. It manages the components inputted by the base-station-side middleware. A group-based approach is used to model a measurement task. A group consists of a master and slaves. The slaves are responsible for measuring designated sensor data and reporting it to the master. The master is responsible for aggregating the data and reporting it to the base station.

Figure 1.7 shows the architecture of the XAC middleware consisting of the base-station-side (XAC Middleware-B) and the sensor-side (XAC Middleware-S).

XAC Middleware-B consists of three components: UserInterface, EventManager and GroupManager. The programmer uses the UserInterface component to register and delete tasks and event handlers, and to obtain the results of the task. The event handler is managed by the EventManager component, and the task is managed by the GroupManager component. The GroupManager component generates Group components corresponding to the task and uses and activates them on the sensor nodes. Each Group component iniates an event when its measurement data satisfies a certain condition and sends it to the EventManager component through the GroupManager. Then, the EventManager component calls the handler corresponding to the event.

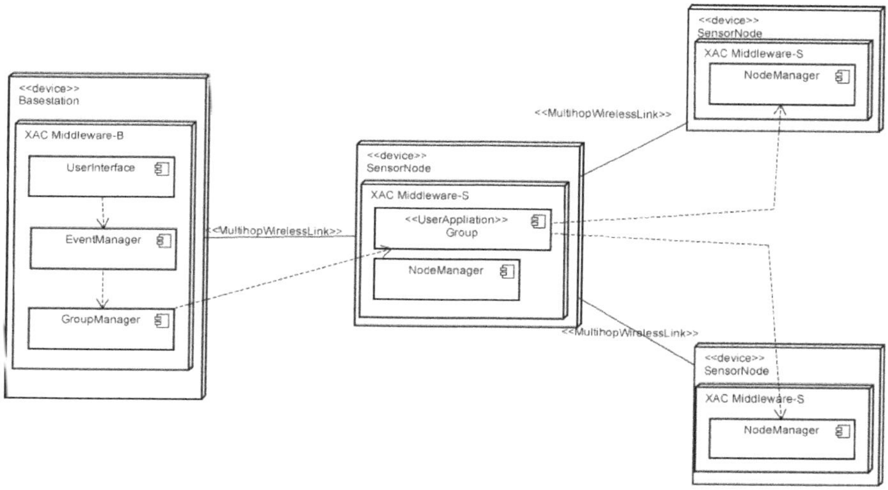

Figure 1.2. *Architecture of prototype implementation*

XAC Middleware-S consists of Group and NodeManager components. The Group has deployment conditions of itself and its measurement tasks. Group components are generated by the GroupManager component of XAC Middleware-B and used on nodes that satisfy its deployment conditions. The Group component finds and selects nodes to satisfy the deployment condition of its measurement task and makes measurement requests of the NodeManager components of those nodes. The Group component collects the measurement results from the NodeManager component and aggregates them. If the aggregated data satisfy the conditions, the Group component fires an event and sends it to the GroupManager component.

The following sections outline the outstanding issues of shared WSN middleware, the existing work on resolving them and the XAC middleware solution from the point of view of the task-description language, runtime task management and self-adaptation.

1.4. Task-description language

Each piece of middleware has its own task-description language. A task-description language is a domain-specific language that specifies the behavior of a WSN for capturing the current context of target phenomena. It provides an abstract view of the WSN to programmers and thus constitutes different levels of programmability.

1.4.1. *Existing solutions*

According to the level of abstraction, existing task-description languages can be classified into data-, group- and node-level languages.

1.4.1.1. *Data-level languages*

Data-level task-description languages, such as those of *TinyDB* [MAD 05], *Cougar* [YAO 02], *TinyLIME* [CUR 05] and *TeenyLIME* [COS 07], allow programmers to describe what kind of data they require and how these data are supposed to be processed to produce a context. These languages each take different approaches to abstraction. For example, *TinyDB* and Cougar abstract the WSN as a relational database and provide SQL-like languages, whereas *TinyLIME* and *TeeneyLIME* abstract the WSN as a tuple space and provide tuple-query languages. Figure 1.2 shows a sample task description written in the language provided by *TinyDB*. The description produces the average of the temperature value measured by all the sensors on the fourth floor of our example environment (section 1.2).

Data-level languages focus not on how to measure data, but on what to measure. Therefore, the concrete behaviors of nodes in the WSN are managed by the middleware and remain transparent to programmers. There are many ways to derive the required data processing functionality. The middleware may support a set of behaviors selected from a potentially adequate functionality. For example, to handle the task description shown in Figure 1.2, *TinyDB* uses a tree-based network topology to route the measured data (i.e. the temperature) and aggregates the data (i.e. to form the average) along the routing tree. However, the programmer cannot change this concrete routing behavior to an alternate one for the purpose of optimization.

```
SELECT AVG(temp)
FROM   sensors
WHERE  floor = 4
SAMPLE PERIOD 5S
```

Figure 1.3. *A task description in TinyDB*

1.4.1.2. *Group-level languages*

Group-level task-description languages, such as those provided by *EnviroTrack* [ABD 04], *Hood* [WHI 04], *Abstract Region* [WEL 04], *Generic Role Assignment* [FRA 05] and *DFuse* [KUM 03], allow the programmer to describe macro-level behaviors for a group of nodes to achieve the desired data processing functionality in a WSN. These languages require a programmer to describe definitions and behaviors of a node group. The programmer usually defines the conditions of the nodes to form a group. *EnviroTrack*, *Hood* and *Abstract Region* form a group consisting of neighboring nodes. A programmer can define which nodes constitute the neighboring ones by using the number of hops from the node closest to the target in the case of *EnviroTrack* or by using physical distances in the case of *Hood* and *Abstract Region*. On the other hand, the programmer may influence the node selection by using node properties such as the battery level, equipped sensor types or bandwidth in the case of *Generic Role Assignment* and *DFuse*.

Moreover, the programmer can define the macro-behaviors of a group by assigning roles to the nodes in a group. For example, the roles used in *EnviroTrack* are classified into two types – the role called *member* measures data and the role called *leader* aggregates or fuses data retrieved from member nodes through a cluster-based network topology. The languages provided by *Hood*, *Abstract Region*, *Generic Role Assignment* and *Dfuse* allow the programmer to describe the network topology between the roles. For example, *Abstract Regions* allow the programmer to select a *geographic*, *planar mesh* or *spanning tree* topology (Figure 1.3). Furthermore, *Dfuse* allows the programmer to define a topology as a data flow graph.

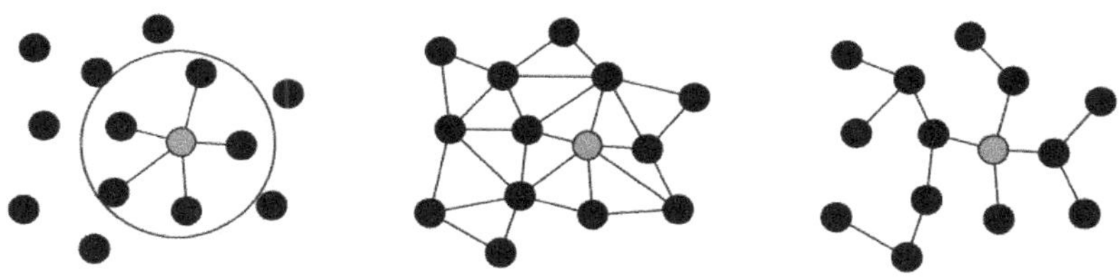

a) Geographic b) Planar mesh c) Spanning tree

Figure 1.4. *Topologies of an Abstract Region*

Compared with data-level languages, group-level languages provide the programmer with a more concrete view of the WSN. Therefore, a task written in a group-level language can be optimized more effectively but this in turn requires detailed knowledge about the node behaviors in the WSN.

1.4.1.3. *Node-level languages*

Middleware such as *Squawk* [SIM 05], *Agilla* [FOK 05], *SensorWare* [BOU 07] and *ActorNet* [KWO 06] provide node-level-description languages. Node-level languages allow a programmer to specify the behavior of a task running on a single node.

Squawk provides the Java programing language to define the behavior of a node. *Agilla*, *SensorWare* and *ActorNet* provide mobile agent-based languages to describe a task that can migrate from one node to another.

Node-level languages provide a more concrete view of WSNs to the programmer than group-level languages do. Since, the behaviors of each node are programmed directly, a task written in a node-level language can be more thoroughly optimized than one written in a group-level language. However, the burden of the programmer increases at the same time. A programmer using a node-level language must have knowledge about distributed programming, since he/she has to implement each task and support, in addition to the main functionality, different aspects like routing, topology management and the synchronization of the nodes.

1.4.2. *XAC middleware solutions*

As indicated above, the existing task-description languages can be classified into data-level, group-level and node-level languages. Each class has different levels of abstraction. Data-level languages provide the most abstract view of a WSN, and node-level languages provide the most concrete view. The level of abstraction constitutes a tradeoff between the description cost and the room for optimization. A higher level of abstraction reduces the description cost, but also reduces the room for optimization at the same time. The adequate level of optimization also depends on the non-functional requirements of the tasks.

The existing solutions each provide only one language with a fixed level of abstraction. For example, *TinyDB* only supports spanning-tree network topologies and does not allow a programmer to use other network topologies such as planar mesh topologies or cluster topologies. From the viewpoint of resource consumption, the selection of an adequate topology depends on the characteristics of the task to be implemented, such as the geographical space to be covered, the type of data to be measured and so on. Therefore, the programmer has to carefully choose an adequate network topology to optimize resource consumption. Group-level or data-level languages allow for this, but data-level languages hide the network topology from the programmer.

XAC middleware provides languages with multiple levels of abstraction. It provides data-level, group-level and node-level languages so that the programmer can choose an adequate level of abstraction. Here, an adequate level of abstraction may depend on the knowledge of the programmer and on the non-functional requirements of his tasks, such as resource consumption, response time or accuracy of the results. Moreover, we are trying to find a proper model transformation for task development (Figure 1.4). A task described at an abstract level should be transformable into a concrete one. A task at the data level may be transformed into a corresponding task at the group level that achieves the data processing functionality described in the data level task with the basic protocols necessary for group management; and a task at the group level may be transformed into a corresponding one at the node level that achieves the macro-behaviors described at the group level.

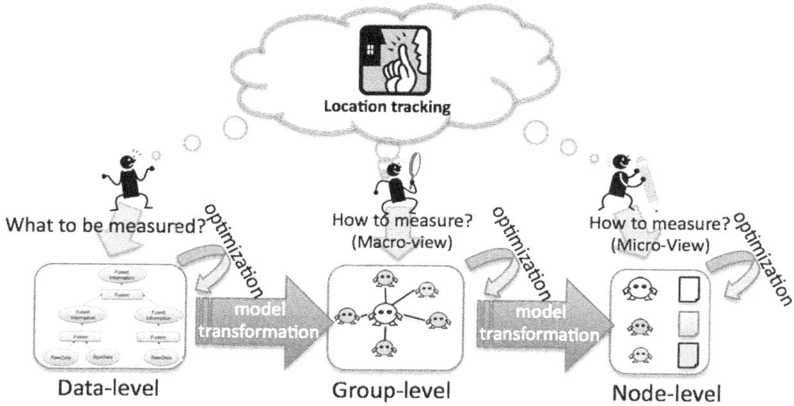

Figure 1.5. *Model-driven development for task descriptions*

A transformed model should have enough information to be executed in the WSN. If needed, the programmer should be able to optimize behaviors at arbitrary levels to fulfill non-functional requirements. We should be able to make suitable optimizations for data processing at the data level, as well as make optimizations of the topology, routing and in-network aggregation in a group at the group level and optimizations of duty-cycling or network device management at the node-level, to achieve the desired accuracy levels, and to decrease resource consumption. For example, consider scenarios S2 and S4 in Table 1.1. Both S2 and S4 require the locations of people near the target location, but S4 requires a higher accuracy than S2. From the viewpoint of data processing, simple localization in which sensors detect a radio signal from beacons that people carry would be enough for S2, but a more sophisticated localization, such as one based on RSSI (radio signal strength indication) that sensors measure, would be required for S4. If such an RSSI-based method is applied at the data processing level, accuracy and resource consumption depend on group-level behaviors such as the number of nodes that measure the RSSI, the routing topology in a group and the in-network aggregation techniques.

The approach presented here is to apply model-driven development to tasks in the WSN. Here, we analyzed the descriptive capabilities of existing languages and constructed a reference model for each level of abstraction. Moreover, we made a catalog of optimization patterns at each level, and one of the transformation patterns from abstract levels to more concrete levels, to support optimization and manual model transformation [TEI 07]. We also constructed transformation rules and development processes according to model transformation and optimization patterns and devised verification methods to ensure consistency between models at each level to guarantee that the behaviors written in higher-level descriptions would still to be valid after subsequent optimizations at lower levels [TEI 14].

1.5. Runtime task management

Sensing tasks have to be assigned to specific nodes for execution, and such resource allocations have to be managed. A rather general approach taken in the early days of wireless sensor networks was to assign tasks to certain nodes by physically connecting each node to a base station and using specific tasks thereby setting up code modules on the nodes. This made changing tasks or assigning new tasks after using the sensor nodes a tedious exercise requiring physical retrieval of the nodes in question. Nowadays, nodes are reprogrammed by assigning tasks dynamically over a wireless communication channel.

1.5.1. *Existing solutions*

Reprogramming techniques can be divided into two distinct approaches:

1) Those managed by the base station: the base station centrally manages tasks and assigns them to the nodes.

2) Those managed by the tasks: the tasks themselves decide which node they use to and move between nodes autonomously[1].

1.5.1.1. *Task deployment management by the base station*

When the base station is used to manage task deployment, it becomes necessary to decide whether tasks should be disseminated throughout the whole network or only sent to specific nodes. For instance, in scenario S1 of Table 1.1, the tasks need only be assigned to nodes inside rooms that need their temperature managed, whereas in S3, the sensing tasks must be used across all the sensor nodes throughout the building. Assuming that changes in, for instance, the sensing conditions and task settings make updating of the tasks of S1 and S3 necessary, we need to consider the range of the affected nodes, namely only for specific nodes as in the case of S1 or in all nodes as in S3.

One existing solution for managing tasks at the base station by specifying a certain range of nodes for updates is Deluge [HUI 04]. Deluge allows the user to optionally define specific node IDs to limit the dissemination of tasks throughout the

1 Tasks can be deployed from the base station instead of moving from one node to another, but they make the deployment decisions on their own in either case.

network. To cope with the energy and communication restraints of wireless sensor nodes, Deluge works to improve efficiency as follows:

– congestion in areas of high node density is avoided by adjusting transmission intervals dynamically;

– asymmetric links are handled by selectively using stable nodes;

– broadcast storms between nodes are avoided by introducing the concept of communication rounds.

Deluge sends the complete code necessary for executing a task from the base station to the nodes. This method is inefficient in so far as it induces high transfer costs that drain the node batteries. To address this issue, other solutions make preliminary deployment of code bases that are shared among different tasks, so that only task-specific updates need to be transferred in case a change becomes necessary. A representative example of this approach is Mate [LEV 02]. Mate provides a virtual machine (VM) on each node that can execute a task-specific lightweight code. Its VM is based on a byte code interpreter that uses 1 byte per byte code. As such, a script consisting of 100 lines amounts only to 100 bytes to be transferred. Mate does not allow the user to specify certain deployment ranges, but the recent revision of Trickle [LEV 04] provides this functionality.

1.5 1.2. *Self-adaptive task deployment management*

In the case that tasks manage themselves autonomously but are used from the base station, the base station is always responsible for distributing tasks to the nodes. In this approach, if it is impossible to preliminarily deploy all tasks to all nodes, changes in the locations in which to execute certain tasks make it necessary to reuse the tasks from the base station. For example, in the previously described scenario S4 in Table 1.1, if the goods to be monitored are moved, the tasks have to be reused to the nodes in the vicinity of the goods' new location every time. This results in a heavy traffic load, especially on nodes around the base station. If the tasks can manage and also reuse themselves by moving autonomously, that burden can be alleviated[2].

To address this problem, the research community has come up with a number of solutions. Representative examples are *Agilla* [FOK 05], *ActorNet* [KWO 06] and *SensorWare* [BOU 03]. *Agilla* extends Mate's VM and gives tasks the ability to move throughout the network. Tasks in *Agilla* are based on a 1–2 byte ISA

2 If the distance between base station and the node where a task is supposed to be redeployed is longer than the distance between the current deployment node and the new deployment location, the cost of moving a task from the current node to the new node is less on average than redeploying the task from the base station.

(instruction set architecture) and thus provide lightweight task codes similar to Mate. *Agilla* can also conduct intra-task communication based on a distributed tuple space [GEL 85].

ActorNet introduces the actor model to wireless sensor networks and executes each actor on its scheme interpreter. It also provides a virtual memory space, scheduling of I/O to the scheme interpreter and garbage collection functionality.

SensorWare is similar to *Agilla* and *ActorNet* in terms of making tasks manage and move themselves. However, it assumes that the deployment platform is an iPAQ and requires about 1 MB of storage and about 128 KB of memory for execution. Thus, its platform specs (several hundred MHz of CPU power and 1 MB of storage) are not in line with those of typical resource-constrained nodes of a wireless sensor networks.

1.5.2. *XAC middleware solutions*

We assume that the wireless sensor network is a shared infrastructure that can be used by any users. Such a network is unlike the traditional ones when it comes to adding, changing and deleting tasks[3].

Furthermore, as described in scenario S4 of Table 1.1, tasks may change their execution location. Consequently, we use task-based autonomous deployment management as discussed in the previous section. The existing solutions allow tasks to be comprised of multiple components running on different nodes. For instance, in the case of scenario S1 of Table 1.1 (temperature management), the task of measuring temperature can be assigned to a certain number of components running on specific nodes inside a room. However, the existing solutions do not provide the relocation of dynamic reconstruction for multiple components. We hence had to come up with our own solutions to provide these functionalities.

1.5.2.1. *Relocation of multiple components*

Our relocation mechanism, called generative dynamic deployment (GDD) of multiple components [SUE 09], provides its functionality through middleware. GDD consists of an architecture to relocate multiple components and a novel relocation method. Past solutions did not offer any architecture for usable components, and thus, the reliability of the relocation process itself was not addressed. Generally speaking, though, if components are to relocate, they need to communicate among each other to coordinate that task, and as a result, the reliability of the relocation task

3 Wireless sensor networks are not capable of preliminary deployment of all tasks for users who suffer from packet loss [SUE 09].

declines with the number of communications. To address this problem, the architecture of GDD comprises three component types (Master, Slave-S and Slave-M), and the Master components delete the Slave-S and Slave-M components only before they relocate. Once they have completed their relocation, they reconstruct the other components on demand.

Figure 1.6 shows an example scenario of relocation based on GDD. The Master, Slave-S and Slave-M components have already been used (1) (not shown) and are about to relocate. The procedure is as follows: (2) the Master component deletes Slave-S and Slave-M components, if relocation becomes necessary; (3) the Master component moves to the new target location; (4) the Master component, upon relocating, reconstructs Slave-S and Slave-M components and uses them on nearby nodes.

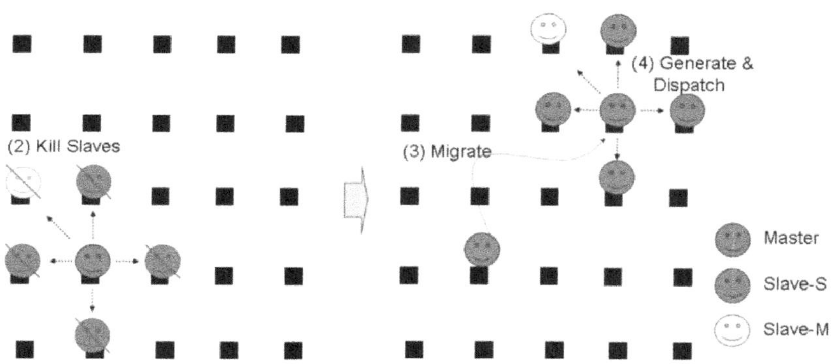

Figure 1.6. *Example scenario based on GDD*

By repeating this process, task execution can be continued at arbitrary locations inside the network. By using GDD, the applicability of tasks comprised of multiple components is increased.

1.5.2.2. *Dynamic components reconstruction*

Dynamic component reconstruction as proposed by Platon *et al.* [PLA 08] also uses tasks as multiple components and adapts the redeployment to the state of the nodes through its Ragilla middleware. In the other existing solutions, the depletion of node batteries or malfunctions of nodes causes tasks comprised of components running on such nodes to fail at fulfilling their requirements. Ragilla, on the other hand, allows the user to define relocation conditions for components (in terms of the number of components, geographical areas, battery level) that control the relocation and reconstruction process in a way that maintains the operational conditions required by each component.

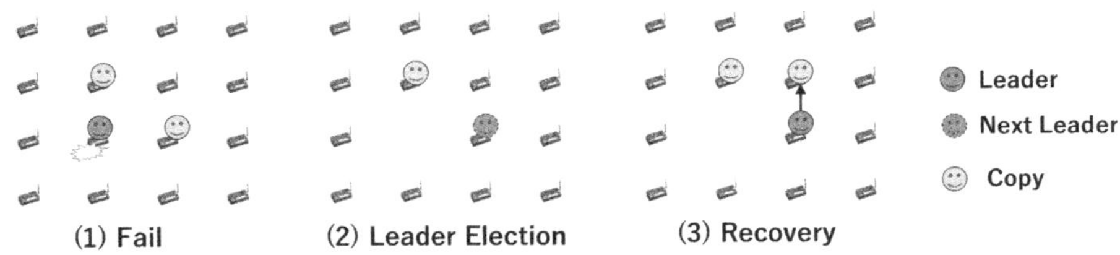

Figure 1.7. *Example scenario based on Ragilla*

For instance, if a hardware failure occurs as illustrated in Figure 1.6, the following steps make it possible to reconstruct the component dynamically at a node within a specified geographical range and continue its sensing task:

1) Because of a hardware failure, a leader component ceases operation.

2) The middleware detects the hardware failure and selects the next leader out of a number of candidate nodes.

3) The middleware also maintains a copy of the leader component and uses it to the selected new node.

By using this method, it becomes possible to improve task execution by reconstructing tasks while maintaining the user's requirements.

Current efforts are not limited to GDD and Ragilla, but they all aim to provide a relocation functionality that adapts dynamically to network conditions, for instance by modeling the role distribution of tasks comprised of multiple components in more detail and making proper deployment decisions based on such component roles.

1.6. Self-adaptation

The situation inside a WSN node, such as the available calculation resource and the battery level, changes over time. Moreover, the situation outside a node, such as the position and velocity of the target and the communication environment, may change as well. In response to these situational changes, the programmer can issue instructions to optimize the behavior of a task. However, if the programmer has to give instructions every time the situation changes, his or her workload would grow too large. To avoid this problem, the middleware should be able to adapt itself to such changes.

1.6.1. *Existing solutions*

1.6.1.1. *Self-adaptability to change inside the node*

Because the processing resource of the node is scarce, it is rather easy for too many tasks to be executed at the same time on the same node. If individual tasks use the nodes of their choice, they would compete for the resources on the nodes and there would come a point when the tasks could not be executed. As described in the section on runtime task management, one task may have to use numerous nodes to, say, improve the quality of an observation. In particular, this means that tasks executed on shared WSNs would frequently compete for the limited resources of the nodes.

There is a certain amount of research dealing with resource competition [HEI 04, DUN 06]. In order to avoid resource competition, each task should make a decision on which nodes to use, and the scheduling of the sensors devoted to the various tasks has to be properly managed.

1.6.1.2. *Self-adaptability to change outside the node*

The task should select the nodes on which it is to run according to the situation. In a pervasive environment, the surroundings of those nodes, such as the presence and number of nearby nodes and the communication links between them, would change dynamically. Moreover, the nodes' surroundings could be affected by other objects. To ensure that the quality level demanded by the task can be met, it becomes necessary to adapt the algorithm and the parameters used by the task in response to the changing situation.

Existing research responds to a changing situation by adjusting the parameters of the communication algorithm [SOH 00, YE 02, SHN 04]. For instance, when a node cannot communicate with another node on the routing tree, these methods restructure the tree so that it can communicate with another node.

1.6.2. *XAC middleware solutions*

1.6.2.1. *Self-adaptive task management*

In the shared WSN, one or more measurement tasks can be used and performed simultaneously, and these tasks can have different accuracy requirements and different priorities. For instance, S2 in Table 1.1 should be used in all rooms, whereas S1 and S3 should only be used in rooms with people in them. The task deployment management should satisfy these requirements as much as possible and sometimes should decide to migrate tasks with lower priorities to other nodes to obey resource constraints. The appropriateness of deployment depends on the current status of the network and/or the current status of the phenomena to be measured.

We devised a self-adaptive algorithm to select the nodes on which to use components for measurements [NAK 08]. The algorithm selects the minimum number of nodes that can satisfy the accuracy requirements of the task, in accordance with the location of the target to be measured and the locations of the sensor nodes. We also devised a self-adaptive algorithm to manage the components in a node [ISH 06, BOU 11]. The component management algorithm evaluates the utility and constraints of the used components and changes the optimal deployment. We formalized this deployment as an integer linear programing problem and used simulated annealing to find semi-optimal deployment within a realistic amount of time.

1.6.2.2. *Self-adaptive communication management*

We also devised a communication algorithm that changes in response to the environment. When the environment around the node changes depending on factors like the movement of the object, the reliability of communication can be improved by selecting the appropriate algorithm for the environment. For instance, in the task of reporting the present positions of employees, reports are sent from nodes in the various rooms they normally work in. However, the reports are sent frequently from the same node when employees gather for a meeting in one room. When the report frequency fluctuates, the middleware can switch the network protocol, such as the routing or MAC protocol, to a more suitable one to reduce communication costs.

Moreover, we developed a self-adaptive algorithm to tune the parameters of the communication algorithm. Communication traffic can be reduced by aggregating data in the group when a result is sent back from the node group. This has prompted research on selecting appropriate nodes and methods of aggregating data. Thanks to this research, we can efficiently perform data aggregation and communication when the routing algorithm changes.

When the environment around the node changes, these can be changed to more appropriate ones for the circumstances, and the quality of the task can be improved.

1.7. Discussion

XAC middleware is useful in three ways to a context management system in a smart city. First, the multi-level task-description language allows users to specify its measurement task without having to know the low level details of a WSN. The users of the context management system are usually interested in sensor data obtainable from the network but may not be experts on wireless *ad hoc* networks. Therefore, we think that our language is suitable for these users. Second, the runtime task management enables ones to add or remove measurements task at runtime. In a smart city, applications will be added or removed at runtime; therefore, the context management system should change their measurement tasks in response to these changes. Finally, the self-adaptation feature reduces unnecessary consumption of resources without the need for human intervention. This feature will prolong the network lifetime and reduce the cost of managing WSNs.

1.8. Conclusion

We described XAC middleware that enables programmers to use a WSN as a smart multi-modal sensor, as part of a context management system for smart cities. It provides different kinds of sensor data to different smart city applications at the same time. A middleware approach is effective at reducing costs associated with developing measurement tasks for a WSN. In this chapter, we overviewed the research issues related to a middleware for a shared WSN, from the viewpoint of task-description language, runtime task management and self-adaptation. We believe that the shared WSN will be a key enabler for smart cities.

1.9 Bibliography

[ABD 04] ABDELZAHER T., BLUM B. *et al.*, "Envirotrack: towards an environmental computing paradigm for distributed sensor networks", *24th International Conference on Distributed Computing Systems (ICDCS)*, pp. 582–5894, 2004.

[BOU 03] BOULIS A., HAN C.-C., SRIVASTAVA M.B., "Design and implementation of a framework for efficient and programmable sensor networks", *1st International Conference on Mobile Systems, Applications and Services (MobiSys)*, pp. 187–200, 2003.

[BOU 07] BOULIS A., HAN C.-C., SHEA R. *et al.*, "Sensorware: programming sensor networks beyond code update and querying", *Pervasive Mobile Computing*, vol. 3, no. 4, pp. 386–412, 2007.

[BOU 11] BOURDENAS T., TEI K., HONIDEN S. *et al.*, "Autonomic role and mission allocation framework for wireless sensor networks", *5th IEEE International Conference on Self-adaptive and Self-Organizing Systems (SASO'11)*, pp. 61–70, 2011.

[COS 07] COSTA P., MOTTOLA L., MURPHY A.L. *et al.*, "Programming wireless sensor networks with the teenylime middleware", *ACM/IFIP/USENIX 2007 International Conference on Middleware (Middleware)*, pp. 429–449, 2007.

[CUR 05] CURINO C., GIANI M., GIORGETTA M. *et al.*, "TinyLime: bridging mobile and sensor networks through middleware", *3rd IEEE International Conference on Pervasive Computing and Communications (PerCom)*, pp. 61–72, March 2005.

[DUN 06] DUNKELS A., FINNE N., ERIKSSON J. *et al.*, "Run-time dynamic linking for reprogramming wireless sensor networks", *4th International Conference on Embedded Networked Sensor Systems (SenSys)*, pp. 15–28, 2006.

[FOK 05] FOK C.-L., ROMAN G.-C., LU C., "Rapid development and flexible deployment of adaptive wireless sensor network applications", *25th IEEE International Conference on Distributed Computing Systems (ICDCS)*, pp. 653–662, 2005.

[FRA 05] FRANK C., ROMER K., "Algorithms for generic role assignment in wireless sensor networks", *3rd International Conference on Embedded Networked Sensor Systems (SenSys)*, pp. 230–242, 2005.

[GEL 85] GELERNTER D., "Generative communication in Linda", *ACM Transaction on Programming Language and Systems*, vol. 7, no. 1, pp. 80–112, 1985.

[HEI 04] HEINZELMAN W.B., MURPHY A.L., CARVALHO H.S. *et al.*, "Middleware to support sensor network applications", *IEEE Network*, vol. 18, no. 1, pp. 6–14, 2004.

[HUI 04] HUI J.W., CULLER D., "The dynamic behavior of a data dissemination protocol for network programming at scale", *2nd International Conference on Embedded Networked Sensor Systems (SenSys)*, pp. 81–94, 2004.

[ISH 06] ISHIGURO M., TEI K., FUKAZAWA Y. *et al.*, "A sensor middleware for lightweight relocatable sensing programs", *International Conference on Computational Intelligence for Modelling, Control and Automation (CIMCA)*, p. 195, 2006.

[KUM 03] KUMAR R., WOLENETZ M., AGARWALLA B. *et al.*, "Dfuse: a framework for distributed data fusion", *1st International Conference on Embedded Networked Sensor Systems (SenSys)*, pp. 114–125, 2003.

[KWO 06] KWON Y., SUNDRESH S., MECHITOV K. *et al.*, "Actornet: an actor platform for wireless sensor networks", *5th International Joint Conference on Autonomous Agents and Multiagent Systems (AAMAS)*, pp. 1297–1300, 2006.

[LEV 02] LEVIS P., CULLER D., "MatÅLe: a tiny virtual machine for sensor networks", *10th International Conference on Architectural Support for Programming Languages and Operating Systems (ASPLOS)*, pp. 85–95, 2002.

[LEV 04] LEVIS P., PATEL N., CULLER D. *et al.*, "Trickle: a self-regulating algorithm for code propagation and maintenance in wireless sensor networks", *1st Conference on Symposium on Networked Systems Design and Implementation (NSDI)*, pp. 15–28, 2004.

[MAD 05] MADDEN S.R., FRANKLIN M.J., HELLERSTEIN J.M. *et al.*, "Tinydb: an acquisitional query processing system for sensor networks", *ACM Transaction on Database Systems*, vol. 30, no. 1, pp. 122–173, 2005.

[NAK 08] NAKAMURA Y., TEI K., FUKAZAWA Y. *et al.*, "Region-based sensor selection for wireless sensor networks", *IEEE International Conference on Sensor Networks, Ubiquitous, and Trustworthy Computing (SUTC)*, pp. 326–331, 2008.

[PLA 08] PLATON E., SUENAGA S., YOSHIOKA N. *et al.*, "Transparent application lifetime management in wireless sensor networks", *Demo Track of the 10th International Conference on Ubiquitous Computing (Ubicomp)*, 2008.

[SHN 04] SHNAYDER V., HEMPSTEAD M., CHEN B.-R. *et al.*, "Simulating the power consumption of large-scale sensor network applications", *2nd International Conference on Embedded Networked Sensor Systems*, pp. 188–200, 2004.

[SIM 05] SIMON D., CIFUENTES C., "The squawk virtual machine: Java on the bare metal", *20th Annual ACM Sigplan Conference on Object-oriented Programming, Systems, Languages, and Applications (OOPSLA)*, pp. 150–151, 2005.

[SOH 00] SOHRABI K., GAO J., AILAWADHI V. *et al.*, "Protocols for self-organization of a wireless sensor network", Personal Communication IEEE, 2000.

[SUE 09] SUENAGA S., YOSHIOKA N., HONIDEN S., "Generative dynamic deployment of multiple components in wireless sensor networks", *6th International Conference on Wireless On-demand Network Systems and Services (WONS)*, pp. 197–204, 2009.

[TEI 07] TEI K., FUKAZAWA Y., HONIDEN S., "Applying design patterns to wireless sensor network programming", *1st International Workshop on Wireless Mesh and ad hoc Networks (WiMAN) in Conjunction with ICCCN*, pp. 1099–1104, 2007.

[TEI 14] TEI K., RYO S., FUKAZAWA Y. *et al.*, "Model-driven-development-based stepwise software development process for wireless sensor networks", *IEEE Transactions on Systems, Man, and Cybernetics: Systems*, vol. 45, no. 4, pp. 675–687, 2014.

[WEL 04] WELSH M., MAINLAND G., "Programming sensor networks using abstract regions", *1st Conference on Symposium on Networked Systems Design and Implementation (NSDI)*, pp. 29–42, 2004.

[WHI 04] WHITEHOUSE K., SHARP C., BREWER E. *et al.*, "Hood: a neighborhood abstraction for sensor networks", *2nd International Conference on Mobile Systems, Applications, and Services (MobiSys)*, pp. 99–110, 2004.

[YAO 02] YAO Y., GEHRKE J., "The cougar approach to in-network query processing in sensor networks", ACM *SIGMOD Record*, vol. 31, no. 3, pp. 9–18, 2002.

[YE 02] YE W., HEIDEMANN J., ESTRIN D., "An energy-efficient mac protocol for wireless sensor networks", *21st Conference on Computer Communications (INFOCOM)*, pp. 1567–1576, 2002.

2

Sensorizer: An Architecture for Regenerating Cyber-physical Data Streams from the Web

The World Wide Web contains a huge amount of data, periodically updated and originally sensed in the physical world. This includes, for example, the density of air pollutants, road traffic conditions and car park occupancy. In many cases, however, this data are not easily accessible from a computer program due to the lack of APIs to fetch them. In this chapter, we propose an architecture for discovering, excavating and streaming the entombed Web content (EWCs) to cope with this problem. This architecture, called Sensorizer, leverages crowd sourcing for accurate EWC discovery, periodic Web scraping with a headless browser for excavation from dynamic webpages and a standardized communication protocol (XMPP) for data streaming for a wide variety of applications.

2.1. Introduction

Smart city applications use the context of the cities they monitor to determine their appropriate behaviors. Context is usually estimated by processing sensor data obtained from sensors used around cities. However, this has a high cost due to the use and management of dozens or hundreds of sensors. Another data source for context estimation would be the World Wide Web. For example, AirNow (http://www.airnow.gov, Figure 2.1) is a Website that discloses current air quality in US cities. Data are acquired from the real world via sensor devices, analysis software, individual interviews and so on. Data are constantly being updated, making past versions of this data totally inaccessible. As a result, cyber-physical data on the Web are continuously entombed.

Chapter written by Jin NAKAZAWA.

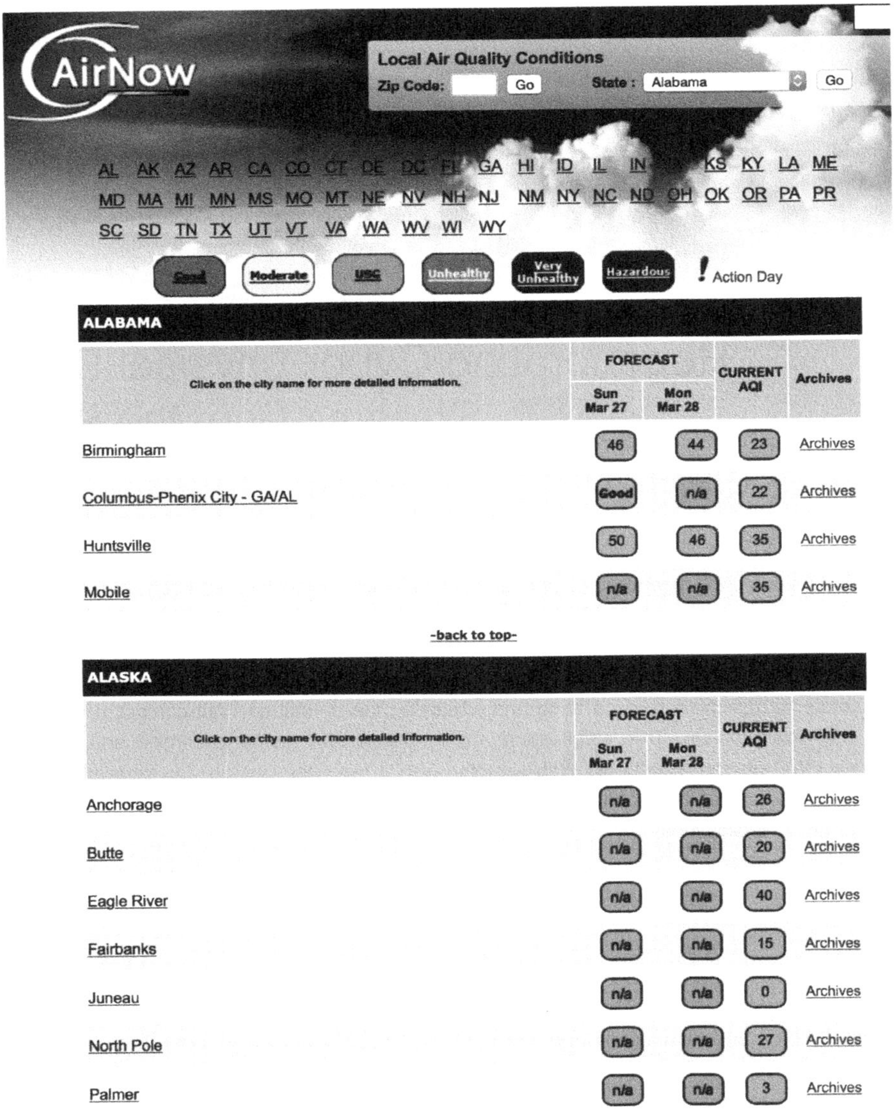

Figure 2.1. *Air quality data entombed online. For a color version of this figure, see www.iste.co.uk/elfallah/smartcities.zip*

In addition, because of the lack of APIs, accessibility to data is low from the application's point of view, although users have access to it. When an application tries to process the air quality values on the AirNow page, it needs to download the whole webpage and scrape the page data. To scrape the webpage, the application needs to have specific knowledge to this website. This difficulty in data consumption results in huge amount of cyber-physical data left unconsumed on the Web, which we have termed entombed web contents (EWCs).

To solve this problem, we have developed a mechanism called Sensorizer, an architecture that regenerates cyber-physical data streams from the Web. It consists of (1) an authoring tool with which an arbitrary element on a webpage can be defined as an EWC container, (2) a probing tool that periodically mines the current value from the container, (3) a data transmission middleware that uses XMPP [SAI 11] over HTTP.

2.2. Sensorizer architecture

In this section we show how EWCs are generated and discuss the design choices to give applications access to them.

2.2.1. *Sensing process of EWC*

In most cases, EWCs are the original data captured from corresponding sensors in the real world. For example, as described above, air quality data is captured from the real world using sensors. In the remainder of the chapter, we refer to any entity which captures data from data sources in the real world as a *sensor*. This phase is called *sensing*. The data are then *collected to a database or alternative storage and analyzed* if necessary. Finally, they are *published* on the Web. If the webpages are machine-friendly, they offer APIs for them to retrieve the data. In most cases, however, they are not. Enabling computer programs to leverage data thus requires an interface somewhere in this process.

2.2.2. *Sensorizer architecture*

Sensorizer, as depicted in Figure 2.2, acts as an interface between EWCs on the Web and computer programs via a set of established technologies. It consists of the browser extension and the probe. As a reference architecture, a minimum of diverse implementations are allowed. Additional functionalities can be introduced for further support for applications. Such functionalities include, for example, storing, searching and processing EWCs.

2.2.2.1. *Browser extension*

Browser extension is a tool that enables users to generate an EWC definition that contains the URL of a webpage with the EWC and its attributes, such as the name, unit, update rate and path. Its architectural role is to crowdsource EWC, finding and associating metadata with them. Leveraging human supervision and knowledge for this allows the architecture for precise excavation of EWC. The browser extension registers the generated EWC definition to the probe using a communication protocol, typically HTTP. This process is what we term *sensorizing*, in which users generate a virtual sensor on the HTML element that contains an EWC.

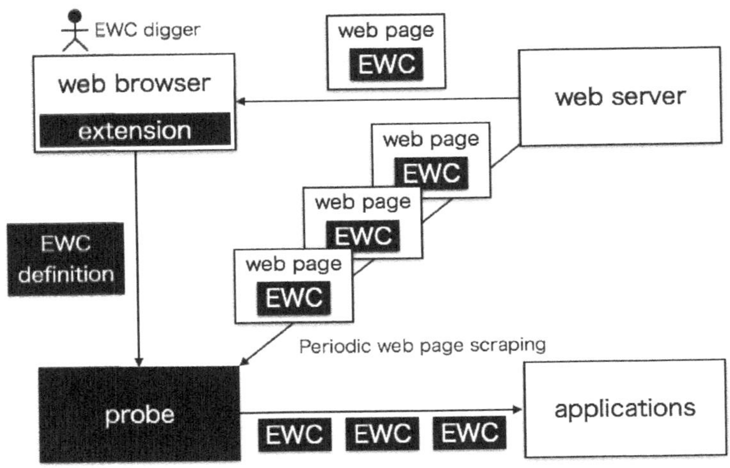

Figure 2.2. *Sensorizer architecture*

EWC definitions registered from a webpage must be shared among users, as it can be seen by multiple users. To do this, the browser extension needs to acquire already registered EWC definitions from the probe, when the page is loaded. Using them, users can know which parts of the page have been already sensorized.

2.2.2.2. *Probe*

A probe periodically downloads the webpages from the URLs that are specified in the registered EWC definitions. It uses the update rate in the definition to determine the rate of webpage downloads. After downloading the pages, it scrapes them to extract the EWC from the path, also specified in the EWC definitions. Finally, it transmits the EWC to a network.

The role of the probe in the architecture is to generate datastreams from a webpage. As discussed above, EWCs are updated periodically. An EWC on a webpage can thus be considered to be a snapshot of the original datastream. The probe allows applications to consume an EWC as a stream, instead of an element packed in an HTML document. It should, however, be noted that temporal granularity in the stream would be coarse, such as data per minute at the finest. As this component fetches webpages continuously from a Web server, it should avoid overloading the server.

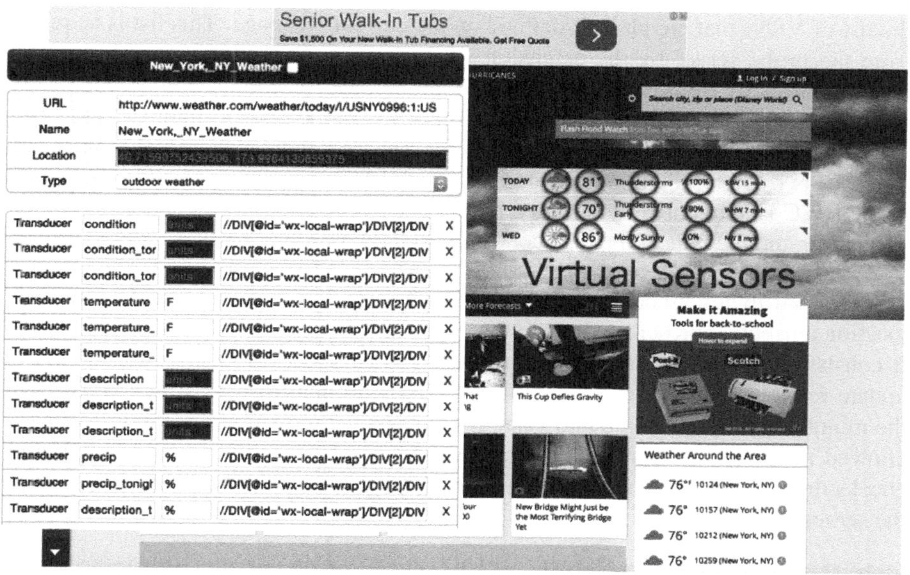

Figure 2.3. *Sensorizer screenshot. For a color version of this figure, see www.iste.co.uk/elfallah/smartcities.zip*

2.3. Implementation

We will now consider a prototype implementation of the *Sensorizer* architecture.

2.3.1. *Sensorizer browser extension*

We have implemented the Sensorizer browser extension to Google Chrome, a screenshot of which is shown in Figure 2.3. The idea of this extension is to let users create virtual sensor nodes (VSNs) on a page. Each VSN can contain multiple virtual transducers, each of which is associated to a single EWC. This allows users to compose multiple data elements related to a single entity into a single virtual node. For example, each row in the table in Figure 2.1 contains information about the air quality in a single city, such as the current AQI and two-day forecast. The screenshot shown in Figure 2.3 depicts the window that shows the virtual sensor node for Birmingham consisting of three transducers, namely `current`, `tomorrow` and `dayAfter`. Doing this allows applications to receive data through a single datastream.

The basic user interaction with this extension is as follows. First, a user can refer to the list of VSNs that are already defined on the current webpage. This list is acquired from the probe server by the extension when the page is loaded. Second, the user creates a VSN by right-clicking the mouse on the page. The extension automatically sets the text content, if any, to the name of the VSN. In the case of Figure 2.1, the user is expected to right-click the mouse on a city. The user then adds virtual transducers to the VSN by right-clicking the mouse on the EWC that he wants to add. The XPath of the EWC is automatically detected by the extension, and retrieves the name of the transducer and the unit of the data contained in the EWC.

Other attributes associated with a VSN include location and update rate. The location attribute represents where in the physical world the EWCs have been sensed. It consists of a latitude and a longitude and can be specified easily on a map. The update rate needs to be specified with the UNIX crontab format, which consists of the minute, hour, day, month and year. This means the maximum scraping frequency allowed in the current Sensorizer implementation is once a minute. Finally, the user checks the radio button on the top left in the pop-up to register the EWC definition in the server.

2.3.2. *Probe*

The probe, invoked periodically by UNIX cron, scrapes EWCs from webpages and transmits them to applications. We implemented the scraping operation with Javascript program run on Phantomjs (http://phantomjs.org), which is so-called a headless Web browser. A headless browser is a fully functional Web browser that does not have graphical user interfaces. This program scrapes EWC, whose paths are specified in the EWC definition. We leverage Phantomjs for the following two reasons. First, it can execute Javascript programs referred within HTML files of a webpage. There exist many webpages that are generated on browsers with Javascript programs. If a VSN definition is generated from such a script-generated webpage, the XPath in the definition is extracted from the DOM structure generated by the script. Scraping meaningful data from such a page, thus, requires the running of scripts to generate the DOM structure equivalent to the above.

Second, Phantomjs completes DOM structure in an HTML file to be fully compliant with HTML. Such a completion functionality is also implemented in Google Chrome on which our browser extension runs. The XPath included in a VSN definition is thus achieved from a completed DOM structure. Using the probe, we can acquire the completed DOM structure again for a successful scraping using Phantomjs.

When sensor data are acquired from a webpage, the probe transmits the data to the Internet using sensor-over-XMPP (SOX) [ROW 11] to achieve higher availability of sensor data over the Internet. XMPP [SAI 11] is an application protocol based on the publish/subscribe model and sensor-over-XMPP specifies sensor data representation using it. The basic idea of SOX is to associate an *event node* in XMPP to a device, a VSN in our case, which contains sensors and/or actuators, which are called transducers.

2.3.3. *Sensorizer/SoX API*

We provide complementary client-side APIs for different programing languages including Java, Javascript and Python on our website (http://sox.ht.sfc.keio.ac.jp). These APIs are object-oriented and share almost the same class hierarchy. The APIs support both publishing and subscription of VSNs.

We also provide Java API for stand-alone applications and server-side Web services. Programmers can leverage the multi-thread capability of the language to handle a number of datastreams with Java API. It also can be used in Java Server Pages (JSP) within Apache Tomcat. In both languages, the XML notation used in XMPP payload is hidden from applications. The developers can thus concentrate on the core body of the programs. In total, we provide support for stand-alone, server-side Web and client-side Web applications consuming EWCs.

2.4. Case of sensorized smart cities

In recent years, smart cities have started to become a reality. The main idea is to sense a wide range of information from the physical world and make it available online. Smart Santander [SCH 11] is one such city where thousands of sensor nodes are embedded to acquire air quality, city noise, parking lot occupancy, users annotation on city locations and so on. SF park (http://sfpark.org) in San Francisco is a sophisticated ecosystem that enables users to obtain parking lot information using a number of dedicated parking meters with wireless communication sensors and APIs for applications to acquire data. These are developed cities in terms of smart city technology.

Many other cities, however, do not have embedded sensors yet. Typically, small pieces of physical information are retrieved from scattered systems, some of which are on webpages. Such pieces of information are, for example, environmental information like those shown in Figure 2.1, traffic information, parking lot occupancy, local city news and many others. Instead of using physical sensors, sensorizing these webpages as follows can help the user to build the information circulation ecosystem.

As we described in the beginning of the chapter, air quality information is available online in many countries. For applications or systems that need access to air quality information, these sites can be their data source. The problem is, however, that these sites are constructed with different structures, whereas the applications' sole interest is air quality. Each application needs to know on these differing webpages, if they directly scrape data from them.

The Sensorizer solves this problem. Once these sites have been sensorized, data on them are published periodically through XMPP. Applications can, thus, consume data by subscribing to the event nodes created by sensorization. They can also find event nodes to subscribe to by using Sensorizer APIs that allow them to look up particular VSNs based on VSN attributes, such as name, location and data unit and/or any keywords included in the webpages where the VSNs were created.

Information on parking lot occupancy is also available on the Web. Figure 2.4 shows a webpage that contains vacancy information on a selected parking lot. The context menu included in this figure is shown when the user right-clicks the mouse over the vacancy icon of the selected parking lot and selects the *Sensorize It-New Device* menu. This results in creating a new VSN on this page. The top-right blue icon is located outside the table structure. Based on our observation on various webpages, EWCs often exist in a similar way in an element outside a constrained structure. Our browser extension lets users find and report EWCs like this, leveraging human eyes and knowledge.

Pages like Highways England (http://www.highways.gov.uk/traffic-information), which show the current traffic information, the number and description of any incidents, change dynamically. This means users cannot determine the number of transducers in a VSN. For such cases, our mechanism is capable of performing a lazy evaluation. Lazy evaluation in our mechanism means that the number of transducers is determined when the page is scraped. To enable this, we allow users to specify the XPath of EWC in a VSN definition with one that can match multiple elements such as `//DIV[@id='events']/TABLE/TBODY/TR/TD[1]`. In this XPath, `TR` match all the `<tr>` elements in the table, so that the complete list of the traffic incidents are always published regardless of the actual number.

Unlike the numeric data above, information on events is described with language and pictures on webpages like local news sites and local government sites.

Sensorizer is capable of probing such data too. For example, *The Times* (http://www.thetimes.co.uk/tto/news/) contains a list of local news in the United Kingdom. By sensorizing such a page, applications can find headlines pushed through XMPP. When a news headline includes a picture, it can also be sensorized. A picture is scraped from a webpage and sent through XMPP with the Data URI Scheme.

Figure 2.4. *Parking lot occupancy in Japan (http://www.repark.jp). For a color version of this figure, see www.iste.co.uk/elfallah/smartcities.zip*

As shown in this section, Sensorizer is capable of generating datastreams from different types of EWC. Applications can integrate this content easily through the common communication protocol, namely XMPP and its XML-based flexible data representation.

2.5. Conclusion

We proposed an architecture for discovering, excavating and streaming entombed Web contents (EWCs). This architecture, called Sensorizer, leverages crowdsourcing for accurate EWC discovery, periodic Web scraping with a headless browser for excavation from dynamic webpages and a standardized communication protocol (XMPP) for data streaming to a wide variety of applications. In this chapter, we first demonstrated the architecture and showed a unique use case, a sensorized smart city, based on sensorizing different webpages and evaluating the architecture. Unlike existing work on web crawler wrapping, which is supposed to work infrequently for updated webpages, our mechanism is capable of generating datastreams from frequently updated ones.

2.6. Bibliography

[ROW 11] ROWE A., BERGES M., BHATIA G. *et al.*, "Sensor Andrew: Large-scale campus-wide sensing and actuation", *IBM Journal of Research and Development*, vol. 55, nos. 1–2, pp. 6:1–6:14, January 2011.

[SAI 11] SAINT-ANDRE P., Extensible Messaging and Presence Protocol (XMPP): Core, Internet RFC 6120, March 2011.

[SCH 11] SCHAFFERS H., SALLSTROM A., PALLOT M. *et al.*, "Integrating Living Labs with Future Internet experimental platforms for co-creating services within Smart Cities", *17th International Conference on Concurrent Enterprising (ICE)*, pp. 1–11, June 2011.

3

Smart Agent Foundations: From Planning to Spatio-temporal Guidance

In this chapter, we highlight how software agents can be conceived to assist users in their tasks and objectives within the smart city. The pertinence of this approach relates to the fact that the agents can react autonomously and on the fly, to the changes of contexts, sometimes unexpected, of their environment. Starting from the common knowledge that intentional belief-desire-intention (BDI) agents can already yield rational behaviours, we embed within each agent a planning process dedicated to the development of an efficient guidance mechanism: (1) the interleavings of the actions coming from the different intention plans are analyzed, highlighting tune solutions to maximize the contextual satisfaction of all the current intentions, taken as a whole, and (2) the past-experiences of the concrete executions of the actions are analyzed to optimize the search of a best solution, in-between performance and duration criteria.

3.1. Introduction

In this chapter, we highlight how software agents can be conceived to assist users in their tasks and objectives within the smart city. These agents can interface the user with the computational world of the smart city. The pertinence of this approach is directly related to the fact that they can react on the fly, to the changes of contexts, sometimes unexpected, of their environment.

The design and development of ambient intelligence (AmI) systems is a promising answer face to the complexity. This involves different challenges in several computing areas as well as intelligent systems research, sensor networks, mobile technologies and interaction within user-centered design [GUI 12, OLA 13].

Chapter written by Ahmed-Chawki CHAOUCHE, Amal EL FALLAH SEGHROUCHNI, Jean-Michel ILIÉ and Djamel Eddine SAÏDOUNI.

From a technical point of view, making some intelligent software agent is a well-known technique for designers of multi-agent systems (MAS), for instance see [RAO 95]. Again, MAS seems to be a powerful approach to design AmI systems. They offer interesting frameworks and support for autonomy, proactivity and societal cooperation needed for such systems. In addition, the belief desire intention (BDI) model [GEO 98] can flexibly handle the entire *modeling* and development of rational agents, based on well-structured and goal-driven aspects, e.g. [CHA 14, CAL 14].

To be context-aware, the major problem for AmI agent consists of recognizing its environmental contexts, including the proper location of the agent and the discovery of other agents and connected objects. Some AmI applications can benefit from a relative common knowledge of the environment, like in demotic application, home centered with some known materials. However, smart cities reveal much more cases [PRE 12]. In such systems, the notion of operational plan of actions become crucial and must be managed on the fly, to cope with the possible changes of context. The recent planning approaches are agent centric rather than handling plans for the whole MAS, actually this eases to embed agent in whatever environment. For instance, the goal – plan structure allows us to handle several alternatives of plans per intentions, and also the management of intentions through plans, e.g. [SAR 06].

In [CHA 14], we proposed a higher order agent (HoA) model offering an explicit separation between management of plans and BDI attitudes. Again, it is agent centric but it comes together with an efficient planning management process, such that the specification of a plan is built directly and automatically from a set of intentions considered by the agent. This answers to a general MAS problem, about the way to achieve some intentions concurrently. The HoA approach is formally based on a formal description language for plans, namely *AgLOTOS*, allowing us to introduce modularity and concurrency aspects to compose sub-plans, viewed as processes.

The planning process, we propose, also aims at offering to each AmI agent a powerful predictive service to guide the agent, contextually and on the fly. Based on an underlying model dedicated to execution of actions and intention plans, a contextual planning system (CPS) was formally demonstrated, which can be exploited to maximize the satisfaction of intentions. Surpassing the results of previous work focusing on the detection of some subset of consistent intentions with respect to some context, [MEN 07], the CPS allows us to compute all the maximum set of consistent intentions, and the way to achieve these intentions concurrently.

Learning techniques are typically used to reinforce the selection of plans to execute. The core idea is to exploit the contextual information encountered previously while running whatever plan. For instance, taking into account the contexts that have allowed the correct performances and the failures of a plan can be used to evaluate the pertinence of the plan to be newly executed in the current context, e.g. [AIR 08]. Closer to our aim, the work of [NUN 14] evaluates the contribution of each plan in terms of utility and preference, seen as parameters for optimization.

In the HoA approach, the learning information is used to reinforce the information of the CPS, from a study of past experiences when executing actions. This leads to a new structure called CPS-L or CPS with learning. Two utility functions are introduced to control the selection, to evaluate the maximum satisfaction solutions either under a safety and/or a speed requirements. Due to the possible unexpected change of context, our learning approach is not based on stochastic processes, such as in [NUN 14], but on an online acquisition of contextual past experiences concerning the executions of actions. In fact, we proceed in two stages, a filtering stage upon the known past experiences to define a pertinent finite subset, then we propose a labeling protocol which balance consideration, between exploration and exploitation. To be efficient, the pertinence of the known past experiences must be analyzed to keep only the pertinent one with respect to the given context. Moreover, as an interesting point, we show how to deal with spatio-temporal information. For instance, it is not really interesting to take the highway at the time it is crowded. This can be captured by a study of the duration of previous transportations via the highway, made at some time during the business day.

The remainder of this chapter is organized as follows: section 3.2 highlights a scenario of an AmI system dedicated to a smart-campus. Section 3.3 describes the functional architecture of our AmI agent. In section 3.4, the HoA model is introduced to capture the evolution of the agent both in its mental and planning states. In section 3.5, we define the AgLOTOS specification language, used to build the agent plan automatically. In section 3.6, we show how the semantics of AgLOTOS is exploited to automatically produce a CPS. Section 3.7 details the acquisition and on the fly exploitation of past experiences regarding the executions of actions. Our spatio-temporal guidance mechanism is then presented, based on the CPS-L structure, consisting of a CPS-structure enriched by the learning information. Section 3.8 concludes our chapter.

3.2. Smart-campus: use case and scenario

We experiment our agent-based approach in a distributed system project called smart-campus. Our aim is to design a powerful system that assists users in their activities within a complex university campus to better interact and adapt to users' needs and expectations.

This proposed application/software offers a useful behavior for agents that allows them to obtain a coherent and stable global behavior by means of only local knowledge, reasoning and actions.

This project is in progress but we concretely equip a float of *Android smart-devices* (SD) by the smart-campus application. In this application, the software architecture is composed of a so-called *HoA agent* and a specific graphical user interface (GUI) to interact with the assisted user. The intrinsic capacities of the HoA agent are developed through the next sections of this chapter. Mainly, it assumes that intelligent software agents can use an efficient guidance service, able to take any perceived change of context on the fly. From a smart-campus architecture, let us now plot the deployment of the smart-campus application in the SD and the way to develop the main processes of the HoA agent.

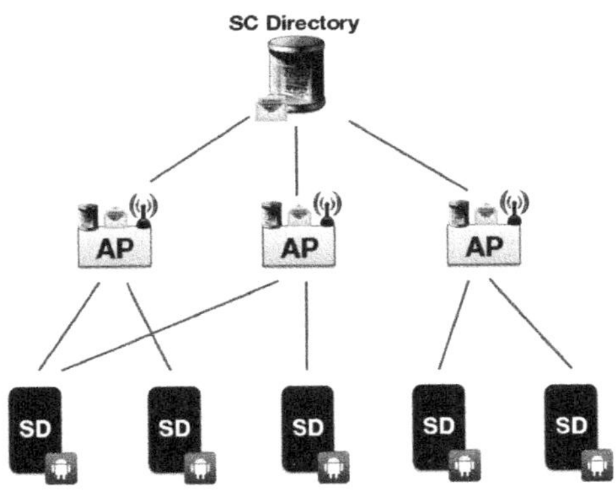

Figure 3.1. *Smart-campus architecture*

3.2.1. *Smart-campus architecture*

The campus system is actualized by the *starting service* of the smart-campus application, which connects the SD to the ``CAMPUS'' network, through one of the possible Wi-Fi access points (AP). As illustrated in Figure 3.1, the SD can automatically access the server, *"SC Directory"*, viewed as a middleware able to maintain the persistence of contextual information like the discovery of users (through their SD) and other connected objects concerning the campus. The starting service is also dedicated to declare the public information of the user at the server level, its location. One of the specificities of this project is that the HoA agent embedded in each SD remains autonomous when the SC directory cannot be reached or when the user is exiting the campus. It can continue assisting the user, due to the perceived context and persistent data previously stored in the SD. Such information can be pervasively updated, by means of the agent's proper capacities or with the help of some neighboring HoA agents.

3.2.2. *Scenario*

The administrative manager of the smart-campus provides a smart application assisting the users through their respective smart devices. We consider two locations in the campus ℓ_1 and ℓ_2 and two users Alice and Bob who can behave on these locations. The desires of Alice are 1) to animate a planned working meeting in ℓ_1 wherein Bob must participate among others, 2) to get her exam copies from ℓ_2. However, she realizes that she cannot make the two following tasks in the same period of time, by herself. In fact, Alice's desires are declared conflictual by her assistive HoA agent, because she is currently in ℓ_1 and does not have enough time to search the copies and come back before the start of the meeting. However, Alice's (assistive) agent proposes an alternative plan consisting in delegating to Bob the task of getting and bringing the copies to her in due time. As the agents are context-aware through the SD, Bob's arrival at the smart-campus can be perceived by Alice's agent. Bob will be asked to bring the copies to the meeting. He can accept that task whose feasibility has been estimated by his (assistive) agent.

3.3. Description of the software architecture for a smart ambient agent

We are interested in modeling the evolution of the agent globally. Figure 3.2 highlights the agent architecture, we consider in this chapter, called HoA architecture. In contrast to the traditional BDI architecture, it underlines a clear separation into four processes:

– *the context process* is in charge of the context information of the agent. It is triggered by new perceptions of the environment and also by internal events informing about the performances of actions. At a low level, it is in charge of observing the realization of the action executions, to state they are successfully achieved or not;

– *the mental process* corresponds to the reasoning part of the agent. It is notified by the context process so that it can be aware of the important context changes and can provoke possible revisions of the beliefs (B), desires (D) and intentions (I) data. Triggered by the perceived events, the mental process updates the B, D and I structures and manages the BDI states of the agent. To organize the selected intentions of the agent, it is able to schedule them by associating with each one a given weight;

– *the planning process* is called by the mental process. Helped by a library of plans (LibP), it mainly produces a plan of actions from the set of weighted intentions. It offers several services, related to the management of plans, like the guidance one which is an intuitive mechanism for assisting the agent in its execution contextually (see section 3.6);

– *the execution process* is called by the planning process. Regarding execution planning, it mainly takes care of the real-time performance of actions and is able to inform us of their successful and failure terminations in plans.

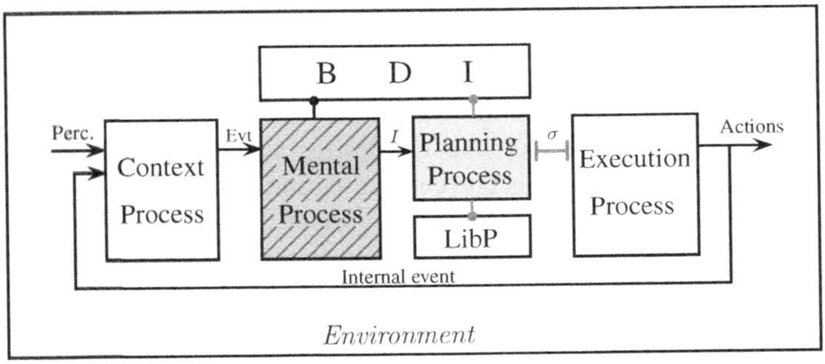

Figure 3.2. *Higher-order agent architecture. For a color version of this figure, see www.iste.co.uk/elfallah/smartcities.zip*

3.4. Higher order agent model

The behavioral model of the agent, called HoA, provides necessary theoretical foundations for designing and studying BDI agents in AmI systems. Based on an *agent-centric* approach, it offers a useful framework to capture the behavior changes of an agent as consequences of its mental evolution and context-awareness ability. At any moment, the current state of an agent is expressed by an *HoA configuration* capturing the evolution of the agent in both 1) its mental reasoning and 2) its planning evolution. Formally, this results in dealing with *HoA configurations* of the agent, as some pairs $q = (bdi, ps)$ where bdi and ps, respectively, represent the BDI and planning states of the agent in q.

As illustrated in Figure 3.3, the evolution of configurations is formally represented by the HoA model, based on an alphabet of events. Events are triggered by the actions being executed or are perception events, $Evt = Evt_{Act} \cup Evt_{Perc}$. Among the actions, message sendings are available, and message receivings are viewed as specific environmental perceptions. Moreover, mobility is handled as a specific action $move$.

DEFINITION 3.1 (The HoA model).– *Typically, a reasoning state of the agent, namely a BDI state, is composed of three sets of propositions, representing the beliefs (B), desires (D) and intentions (I). Furthermore, we will see them globally and simply speak about $\mathcal{BDI}$ for the set of all the possible states that can be defined over the BDI structure of the agent. To define the HoA model, let also introduce $\mathcal{P}$ as the set of all the possible corresponding plans of the agent, let $\mathcal{PS}$ be the set of all the possible* *planning states evolving in some plan, and let $LibP$ be a subset of $\mathcal{P}$ representing the library of plans. The HoA model of the agent is a transition system, represented by a tuple $\langle Q, q_0, \rightarrow, F_M, F_P, F_{PS} \rangle$, where:*

– Q is the set of HoA configurations;

– $q_0 \in Q$ is the initial HoA configuration;

– is the set of transitions between configurations;

– $F_M : Q \longrightarrow \mathcal{BDI}$ associates a BDI state with each HoA configuration;

– $F_P : \mathcal{BDI} \times \; LibP \longrightarrow \mathcal{P}$ associates with each BDI state, an agent plan built from the LibP library;

– $F_{PS} : Q \longrightarrow \mathcal{PS}$ associates a planning state with each HoA configuration.

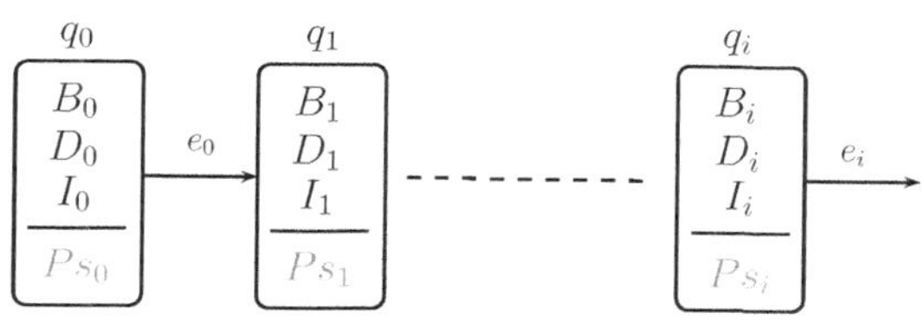

Figure 3.3. *The agent behavioral changes. For a color version of this figure, see www.iste.co.uk/elfallah/smartcities.zip*

The reader will see further how at any moment, the BDI state and the planning state of any reached HoA configuration of the agent are consistent. Roughly speaking, we consider that the BDI reasoning of the agent is able to deliver a set of intentions that is partially ordered. Such an ordering is useful to solve possible conflicts that could exist contextually between the intentions of the agent. In section 3.5, we show how to directly deduce the corresponding plan from any weighted set of intentions. It takes the form of an AgLOTOS expression. As AgLOTOS is a formal language, the possible consistent planning states are derived by using the semantics of AgLOTOS contextual. More formal details are given in section 3.6, but let us first introduce a frame scenario to follow the explanations.

3.4.1. *Application to the scenario*

Consider the scenario of section 3.2 again. Table 3.1 separately represents a possible evolution of HoA configurations for the agent Alice and Bob, to solve the Alice's problem. For the sake of clarity, only the evolution of the BDI parts of HoA configurations is represented, but complementary details concerning the associated planning state evolutions are presented further in Table 3.6.

Alice's scenario	
q_0^A	$B_0 = \{in(me,\ell_1), in(copies,\ell_2)\}$ $D_0 = \{meeting(Bob,\ell_1), getting\_copies(\ell_2)\}$ $I_0 = \{meeting(Bob,\ell_1)\}$
q_1^A	$B_1 = \{in(me,\ell_1), in(copies,\ell_2), in(Bob,\ell_2)\}$ $D_1 = \{meeting(Bob,\ell_1), asking(Bob, get\_copies(\ell_2))\}$ $I_1 = \{meeting(Bob,\ell_1), asking(Bob, get\_copies(\ell_2))\}$

Bob's scenario	
q_0^B	$B_0 = \{in(me,\ell_2)\}$ $D_0 = \{waiting(\nu), meeting(Alice,\ell_1)\}$ $I_0 = \{waiting(\nu), meeting(Alice,\ell_1)\}$
q_1^B	$B_1 = \{in(me,\ell_2), in(copies,\ell_2)\}$ $D_1 = \{meeting(Alice,\ell_1), getting\_copies(\ell_2)\}$ $I_1 = \{meeting(Alice,\ell_1), getting\_copies(\ell_2)\}$

Table 3.1. *A state evolution for Alice and Bob*

To express the agent plan configurations, the BDI propositions and plan of actions are simply expressed by using instantiated predicates, e.g. $getting\_copies(\ell_2)$. The initial configurations of Alice and Bob agents are respectively q_0^A and q_0^B.

The proposed problem of Alice is that she cannot process the two following tasks at the same time: 1) to meet Bob in ℓ_1, 2) to get her exam copies from ℓ_2. After having perceived that Bob is in ℓ_2 ($perc(in(Bob,\ell_2))$), meaning in the same location as the exam copies, Alice's agent enriches its beliefs, desires and intentions, communicating with Bob's agent and asking for help to bring her the copies.

At this point, this mental process can be assisted by a guidance service which confirms that Bob can satisfy both desires consistently, to get the copies and to meet Alice, taking into account the belief ($in(copies,\ell_2)$). The mental process of Bob's agent then considers the additional intention $getting\_copies(\ell_2)$. Hence, the HoA configuration of Bob's agent is changes to q_1^B, to specify both desires.

Next, the AgLOTOS language is used to express how the concurrent plan of Bob's agent is formally built, according to its intentions.

3.5. Description of the concurrent planner based on AgLOTOS language

3.5.1. *Agent plan structure*

Starting from any BDI state, we now show how to build an *agent plan* ($\overline{P}$) automatically. The library LibP of the operational plans is used as a basis to build it.

The plans in the library are called *elementary plans* (P) and are indexed by the possible intentions known at the agent level. As highlighted in Figure 3.4, a plan of a given intention, called an *intention plan* ($\widehat{P}$), is often an alternate of several elementary plans where each one can satisfy the corresponding intention. An agent plan agrees with an *agent plan structure* expressing all the capacities of elementary plan alternations and of concurrency to achieve intentions. Table 3.2 describes the notations used in this approach.

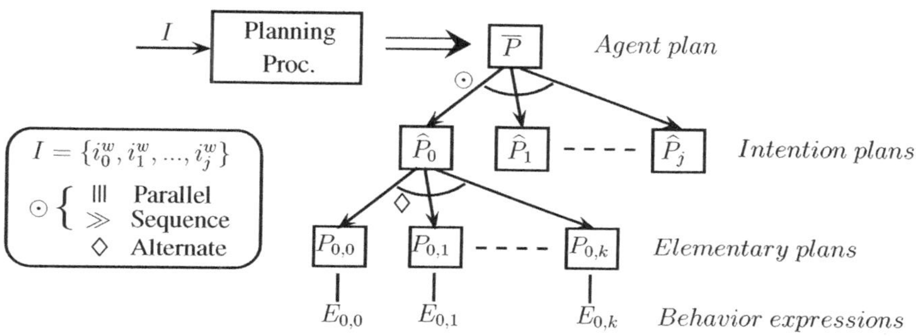

Figure 3.4. *Agent plan structure*

Notation	Description
q	HoA configuration
bdi	BDI state
ps	Planning state
E	AgLOTOS expression
P	Elementary plan
$\widehat{P}$	Intention plan
$\overline{P}$	Agent plan
(E, P)	Elementary plan configuration
$(E, \widehat{P})$	Intention plan configuration
$[\overline{P}]$	Agent plan configuration

Table 3.2. *Synthetic presentation of the used notations*

The proposed agent plan structure is formally captured by using a formal description language, namely AgLOTOS [CHA 14]. AgLOTOS inherits from the LOTOS language [BRI 88] so offers different ways to express the concurrency of actions and the modularity of plans.

3.5.2. *Syntax of AgLOTOS plans*

3.5.2.1. *Agent plan level*

The set of intention plans can be globally composed by using the concurrent $|||$ or sequential $\gg$ operators, leading to the specification of an agent plan. Let $\overline{\mathcal{P}}$ be the set of names qualifying the possible agent plans with $\overline{P} \in \overline{\mathcal{P}}$, and $\widehat{\mathcal{P}}$ be the set of names used to identify the possible intention plans with $\widehat{P} \in \widehat{\mathcal{P}}$, such that $\overline{P}$ is any agent plan defined by

$$\overline{P} ::= \widehat{P} \mid \overline{P} ||| \overline{P} \mid \overline{P} \gg \overline{P}$$

3.5.2.2. *Intention plan level*

With respect to an intention, the alternation of elementary plans is described in AgLOTOS by using the composition operator $\Diamond$. An intention is satisfied if and only if at least one of the associated elementary plans is successfully executed. Formally, let $\mathcal{P}$ be the set of names qualifying the possible elementary plans with $P \in \mathcal{P}$. We define an intention plan $\widehat{P}$ as

$$\widehat{P} ::= P \mid \widehat{P} \Diamond \widehat{P}$$

3.5.2.3. *Elementary plan level*

Any elementary plan is described by an *AgLOTOS expression*, referring to a finite set of observable actions. Let $\mathcal{O}$ be the (finite) set of observable actions which are viewed as instantiated predicates, ranging over $a, b, ...,$ and let L be any subset of $\mathcal{O}$. $H \subset \mathcal{O}$ is the set of the so-called mobility and communication primitives. Let $Act = \mathcal{O} \cup \{\tau\}$, be the set of actions, where τ is the internal action.

– In AgLOTOS, actions are refined to make the AmI primitives observable: 1) an agent can communicate with another agent in the AmI system, 2) it can perceive its enter and leave in the system, 3) it can move between the system locations.

– Any AgLOTOS expression is associated with contextual information relating to the (current) BDI state of an agent: 1) Θ is a finite set of space locations where the agent can move, 2) Λ is a set of agents with which it is possible to communicate, 3) $\mathcal{M}$ is the set of possible messages to be sent and received.

– The agent mobility is expressed by the primitive $move(\ell)$ which is used to handle a move of the agent to some location ℓ ($\ell \in \Theta$). The syntax of communication primitives is inspired from the semantics of π-calculus primitives, however with the consideration of a totally dynamic communication support, hence without specification of predefined channels: the expression $x!(\nu)$ specifies the emission to the agent x ($x \in \Lambda$) of some message ν ($\nu \in \mathcal{M}$), whereas the expression $x?(\nu)$ means that ν is received from some agent x.

The syntax of an elementary plan P is defined inductively as follows:

$$
\begin{array}{lll}
P ::= & E & \\
E ::= & exit \mid stop & \\
 & \mid a; E \mid E \odot E & (a \in \mathcal{O}) \\
 & \mid hide\ L\ in\ E & \\
H ::= & move(\ell) & (H \subset \mathcal{O}, \ell \in \Theta) \\
 & \mid x!(\nu) \mid x?(\nu) & (x \in \Lambda) \\
\odot = & \{ [\,], \gg, [>, |[L]|, ||, ||| \} &
\end{array}
$$

In this syntax, $P ::= E$ represents an elementary plan identified by P, such that its behavior expression is E. The elementary expression $stop$ specifies a plan behavior without possible evolution and $exit$ represents the successful termination of some plan. The expression $a; E$ denotes an action a prefixing E, and the set $\odot$ represents the standard LOTOS operators and some of them can refer to any subset L of $\mathcal{O}$. The expression $E [\ ] E$ specifies a non-deterministic choice, $E \gg E$ a sequential composition and $E\ [>\ E$ the interruption. The LOTOS parallel composition, denoted by $E\,|[L]|\,E$, can model *both synchronous* composition for actions defined in L, denoted by $E\,||\,E$ with $L = \mathcal{O}$, and *asynchronous composition*, denoted by $E\,|||\,E$ with $L = \emptyset$. The expression $hide\ L\ in\ E$ represents an explicit hiding of actions mentioned in L, making them unobservable in E.

Every expression of a process is terminated by an $exit$. Here are now different examples of compositions between two sub-expressions E_1 and E_2, with $E_1 ::= a; b; exit$ and $E_2 ::= a; b; c; exit$:

– $E ::= E_1 \gg E_2$ means that all the expression E_1 is realized before the expression E_2;

– $E ::= E_1\ |||\ E_2$ means that the expression E_1 and E_2 are realized in concurrence;

– $E ::= E_1\ [\,]\ E_2$ means that only one of E_1 and E_2 is realized;

– $E ::= E_1\ [>\ E_2$ means that the expression E_1 can no longer be executed when the execution of E_2 starts;

– $E ::= E_1\ |[b]|\ E_2$ means that the joint execution of b in E_1 and E_2 is possible only when b is possible in both E_1 and E_2;

– $E ::= hide\ a\ in\ (E_1\ |[b]|\ E_2)$ means that the action a is hidden when a is offered in the expression $(E_1\ |[b]|\ E_2)$. In fact, this operator makes all the occurrences of the action a unobservable (replaced by τ) at the semantical level (see section 3.6.1).

3.5.3. *Building of the agent plan from the intentions*

With respect to the set of intentions I of the agent, the agent plan is formed in two steps: 1) by an extraction mechanism of elementary plans from the library LibP and 2) by using the composition functions called $options$ and $plan$:

– $libp : \mathcal{I} \rightarrow 2^{\mathcal{P}}$ features the library of elementary plans. It yields, for each intention $i \in \mathcal{I}$, a set of instantiated elementary plans dedicated for achieving i;

– $options : \mathcal{I} \rightarrow \widehat{\mathcal{P}}$ yields for any $i \in \mathcal{I}$, an intention plan of the form: $\widehat{P_i} = \Diamond_{P \in libp(i)} P$;

– creates the final agent plan $\overline{P}$ according to the set of intentions I. Depending on how I is ordered, the intention plans yielded by the different mappings $\widehat{P_i} = options(i)$ such that $i \in I$ are composed by using the composition operators $|||$ and $\gg$.

To account for agent intentions, we propose that the agent can label the different elements of the set I of intentions by using a weight function $weight : I \rightarrow \mathbb{N}$. The ones having the same weight are composed by using the parallel operator $|||$. In contrast, the intention plans corresponding to distinct weights are ordered by using the sequential operator $\gg$.

For instance, let $I = \{i_g^2, i_e^1, i_m^2\}$ be the considered set of intentions, such that the superscript information denotes a weight value. As illustrated in Figure 3.5, let $\widehat{P_g}, \widehat{P_e}, \widehat{P_m}$ be their corresponding intention plans, the constructed agent plan could be viewed as $\overline{P} ::= (\widehat{P_g} ||| \widehat{P_m}) \gg \widehat{P_e}$, where $\widehat{P_g} ::= P_g$, $\widehat{P_m} ::= P_{m1} \Diamond P_{m2} \Diamond P_{m3}$ and $\widehat{P_e} ::= P_{e1} \Diamond P_{e2}$. Furthermore, each elementary plan P_k is associated with an expression E_k describing its behavior, as defined in section 3.5.2.

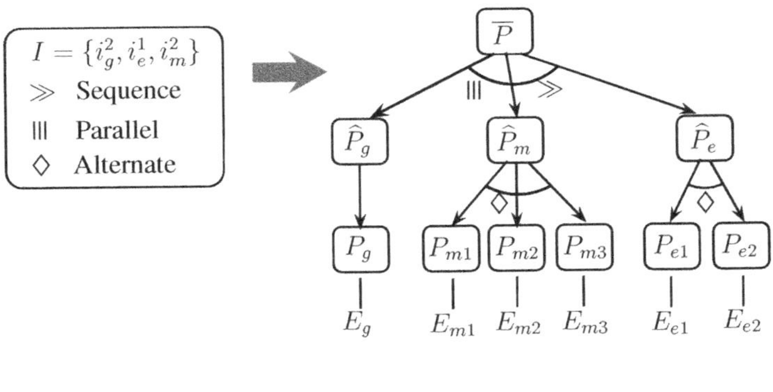

Figure 3.5. *An example of an agent plan structure*

3.5.4. *Planning state of the agent*

The state of an agent plan $\overline{P}$, also called an *agent plan configuration*, is denoted by $[\overline{P}]$. From the plan structure of any agent plan, such configuration is described by associating a behavior expression with each elementary plan. and by composing them according to the used algebraic operators. The *canonical rules* of definition 3.2 specify how $[\overline{P}]$ is formed compositionally from some *intention plan configurations*, like $(E, \widehat{P})$ (*rule 1*), themselves built from an alternate of *elementary plan configurations*, like (E_k, P_k) (*rule 2*) which represents an elementary plan identified by P_k, and its behavior expression is E_k.

Considering the former example, $[\overline{P}] = \Big((E_g, \widehat{P_g})|||(E_{m1} \Diamond E_{m2} \Diamond E_{m3}, \widehat{P_m})\Big) \gg (E_{e1} \Diamond E_{e2}, \widehat{P_e})$.

DEFINITION 3.2 (Generic representation of an agent plan configuration).– *Any agent plan configuration $[\overline{P}]$ has a canonical representation defined by the following two rules:*

– 1) $$\frac{\overline{P}::=\widehat{P} \qquad \widehat{P}::=\Diamond^{k=1..n}\, P_k \qquad P_k::=E_k}{[\overline{P}]::=(\Diamond^{k=1..n}\, E_k,\ \widehat{P})}$$

– 2) $$\frac{\overline{P}::=\overline{P_1} \odot \overline{P_2} \qquad \odot\in\{|||,\gg\}}{[\overline{P}]::=[\overline{P_1}] \odot [\overline{P_2}]}$$

Be aware that an agent plan configuration defined so can be considered as an initial planning state. By applying the operational semantics of AgLOTOS, different agent plan configurations can be derived from this configuration, as described in the next section.

3.6. Contextual planning guidance

In this section, we show how to build a contextual planning system (CPS). It is a transition system representing all the possible evolutions of the agent plan in terms of actions. With respect to some evolution, a *contextual planning state* takes into account a spatial information, precisely the agent location, and also an information on the intention plans to be achieved.

DEFINITION 3.3 (Contextual planning state).– *A contextual planning state is a tuple (ps, ℓ, T), where ps is any planning state, ℓ corresponds to a contextual information, here restricted to a location for the agent and T is the subset of intention plans which are terminated.*

The building of CPS transitions is formally driven by the semantics of AgLOTOS. From any contextual planning state, the semantics yields the actions that can be offered and also, for each one, the resulting contextual planning states. The semantics is composed of three levels of derivation rules, which are presented separately in Tables 3.3, 3.4 and 3.5. Semantically, these levels are nested: the rules of the agent plan level are refined by the ones of the intention plan level, themselves refined by the ones of the elementary plan level.

3.6.1. *Semantics of AgLOTOS plans*

3.6.1.1. *Elementary plan level*

The semantic rules defining the possible changes of an elementary plan configuration are detailed in Table 3.3. In particular, the *(Plan definition) rule* specifies how a pair (E, P) is changed to (E', P) under some action a. Actually, $P := E$ means to consider any (E, P) source pair and $P \xrightarrow{a} E'$ means changing E to E' for P under the execution of a.

For instance, if we consider the elementary plan configuration (E_g, P_g) such that $E_g = getc; confirm; exit$, then the action $getc$ can be offered to the execution, i.e. $P_g \xrightarrow{getc} E'_g$. This results in the configuration (E'_g, P_g) such that $E'_g = confirm; exit$.

3.6.1.2. *Intention plan level*

When an intention plan is assumed to be treated, a transition $(ps_1, a, \widehat{P}, ps_2)$, denoted by $ps_1 \xrightarrow[\widehat{P}]{a} ps_2$, expresses a change of intention configuration, from ps_1 to ps_2. In Table 3.4, the fist rule assumes the execution of the action a from $E \xrightarrow{a} E'$ and $P := E$, whereas the second one highlights the termination case, keeping trace of the intention plan $\widehat{P}$ that is going to be terminated. By calling PS, the set of all the possible intention plan configurations for the agent, the transition relation is a subset of $PS \times Act \times \widehat{P} \times PS$. For sake of clarity, the transition (ps_1, a, nil, ps_2) is simply denoted by $ps_1 \xrightarrow{a} ps_2$.

Observe that due to the fact we consider a predictive guidance in the CPS, only successful executions are taken into account, thus abstracting that a plan may fail. Moreover, the semantics of the alternate operator is reduced to the simple non-deterministic choice of LOTOS: $\Diamond^{k=1..n} E_k \equiv [\,]^{k=1..n} E_k$ to take into account every elementary plan for achieving the corresponding intention.

3.6.1.3. *Agent plan level*

The possible changes of contextual planning states, like (ps, ℓ, T), are expressed in Table 3.5. In *(Action) rules*, the first rule shows the case of a regular action, whereas the second one specifies the termination case of some intention plan, which is added to T. In *(Communication) rules*, the action send $x!(\nu)$ (respectively receive $x?(\nu)$) is constrained by the visibility of the agent x in its neighborhood. In *(Mobility) rules*, the effect of the action $move(\ell')$ yields the agent to be placed in ℓ'.

(Termination)	$\dfrac{}{exit \xrightarrow{\delta} stop}$
(Action prefix)	$\dfrac{a \in \mathcal{O}}{a;E \xrightarrow{a} E}$
(Choice)	$\dfrac{E \xrightarrow{a} E'}{F \; [] \; E \xrightarrow{a} E' \qquad E \; [] \; F \xrightarrow{a} E'}$
(Concurrency)	$\dfrac{E \xrightarrow{a} E' \qquad a \notin L \cup \{\delta\}}{E \; \vert[L]\vert \; F \xrightarrow{a} E' \; \vert[L]\vert \; F}$
	$\dfrac{E \xrightarrow{a} E' \qquad a \notin L \cup \{\delta\}}{F \; \vert[L]\vert \; E \xrightarrow{a} F \; \vert[L]\vert \; E'}$
	$\dfrac{E \xrightarrow{a} E' \quad F \xrightarrow{a} F' \quad a \in L \cup \{\delta\}}{E \; \vert[L]\vert \; F \xrightarrow{a} E' \; \vert[L]\vert \; F'}$
(Hiding)	$\dfrac{E \xrightarrow{a} E' \qquad a \notin L}{hide \; L \; in \; E \xrightarrow{a} hide \; L \; in \; E'}$
	$\dfrac{E \xrightarrow{a} E' \qquad a \in L}{hide \; L \; in \; E \xrightarrow{\tau} hide \; L \; in \; E'}$
(Sequence)	$\dfrac{E \xrightarrow{a} E' \qquad a \neq \delta}{E \gg F \xrightarrow{a} E' \gg F}$
	$\dfrac{E \xrightarrow{\delta} E'}{E \gg F \xrightarrow{\tau} F}$
(Interruption)	$\dfrac{E \xrightarrow{a} E' \qquad a \neq \delta}{E \; [> F \xrightarrow{a} E' \; [> F}$
	$\dfrac{E \xrightarrow{\delta} E'}{E \; [> F \xrightarrow{\delta} E'}$
	$\dfrac{F \xrightarrow{a} F'}{E \; [> F \xrightarrow{a} F'}$
(Relabeling)	$\dfrac{E \xrightarrow{a} E' \qquad a \notin \{a_1,\ldots,a_n\}}{E[b_1/a_1,\ldots,b_n/a_n] \xrightarrow{a} E'[b_1/a_1,\ldots,b_n/a_n]}$
	$\dfrac{E \xrightarrow{a} E' \qquad a = a_k \; (1 \leq k \leq n)}{E[b_1/a_1,\ldots,b_n/a_n] \xrightarrow{b_k} E'[b_1/a_1,\ldots,b_n/a_n]}$
(Plan definition)	$\dfrac{P := E \quad E \xrightarrow{a} E'}{P \xrightarrow{a} E'}$

Table 3.3. *Semantic rules of elementary plan configurations*

(Action)	$\dfrac{E \xrightarrow{a} E' \quad a \in Act}{(E,\widehat{P}) \xrightarrow{a} (E',\widehat{P})}$
	$\dfrac{E \xrightarrow{\delta} E' \quad E = exit}{(E,\widehat{P}) \xrightarrow[\widehat{P}]{\tau} (E',\widehat{P})}$

Table 3.4. *Semantic rules of intention plan configurations*

3.6.2. *Contextual planning system*

From any intention set of the agents, denoted by I, a CPS is built by using the semantic rules of AgLOTOS.

DEFINITION 3.4 (Contextual Planning System).– *The CPS is a labeled Kripke structure* $\langle S, s_0, Tr, \mathcal{L}, \mathcal{T} \rangle$ *where:*

– S *is the set of contextual planning states;*

– $s_0 = (ps, \ell, \emptyset) \in S$ *is the initial contextual planning state of the agent;*

– $Tr \subseteq S \times Act \times S$ *is the set of transitions. The transitions are denoted by* $s \xrightarrow{a} s'$ *such that* $s, s' \in S$ *and* $a \in Act$*;*

– $\mathcal{L} : S \to \Theta$ *is the location labeling function;*

– $\mathcal{T} : S \to 2^{\widehat{\mathcal{P}}}$ *is the termination labeling function which captures the terminated intention plans.*

The CPS is built from an initial contextual planning state, e.g. $([\overline{P}], \ell, \emptyset)$ such that $[\overline{P}]$ is the initial agent plan configuration, and ℓ is the location ℓ currently considered for the agent. At that point, all the intention plans mentioned in $[\overline{P}]$ are specified not being started (i.e. $T = \emptyset$).

In a CPS, any transition $s \xrightarrow{a} s'$ represents an action to be performed. Like in the STRIPS description language [MEN 07], the actions are associated with preconditions and effects. In our approach, the preconditions only concern the contextual information attached to the source state. Let $pre(a)$ be the precondition of any action a, e.g. $pre(a(\ell)) = \ell = \mathcal{L}(s)$. In Figure 3.6, the three actions that are not realizable are represented by dashed transitions from the states s_2, s_5 and s_9. From these states, $pre(getc(\ell_2)) = \ell_2 \neq \mathcal{L}(s)$.

To guide the agent efficiently, the planning process can select an execution trace which maximizes the number of intentions that can be achieved. This can be captured over the set $\Sigma \subseteq 2^{Tr}$ of all the possible traces of the CPS. We introduce the notion of *maximum trace* based on the mapping $end : \Sigma \longrightarrow 2^{\widehat{\mathcal{P}}}$, used to specify the set

$end(\sigma)$ of the different terminated intention plans that occur in a trace $\sigma \in \Sigma$. Let Σ_{MAX} represent the set of maximum traces of the CPS. As a specific tree structure, Figure 3.10 represents the 10 maximum traces of CPS_B. In this unfolded version of Figure 3.6, the end set of a trace σ corresponds to the termination set attached to the final state of σ, given by the semantics of the CPS. Moreover, the trace carried out by $s_0 \to s_2 \to s_5 \to s_9$ in Figure 3.6 is not represented because it is not a maximum trace.

(Action)	$\dfrac{ps \xrightarrow{a} ps' \quad a \in Act}{(ps,\ell,T) \xrightarrow{a} (ps',\ell,T)}$
	$\dfrac{ps \xrightarrow[\widehat{P}]{\tau} ps'}{(ps,\ell,T) \xrightarrow{\tau} (ps',\ell,T \cup \{\widehat{P}\})}$
(Communication)	$\dfrac{ps \xrightarrow{x!(\nu)} ps' \quad x \in \Lambda}{(ps,\ell,T) \xrightarrow{x!(\nu)} (ps',\ell,T)}$
	$\dfrac{ps \xrightarrow{x?(\nu)} ps' \quad x \in \Lambda}{(ps,\ell,T) \xrightarrow{x?(\nu)} (ps',\ell,T)}$
(Mobility)	$\dfrac{ps \xrightarrow{move(\ell')} ps' \quad \ell \neq \ell'}{(ps,\ell,T) \xrightarrow{move(\ell')} (ps',\ell',T)}$
	$\dfrac{ps \xrightarrow{move(\ell)} ps'}{(ps,\ell,T) \xrightarrow{\tau} (ps',\ell,T)}$
(Sequence)	$\dfrac{ps_1 \xrightarrow{a} ps'_1 \quad a \in Act}{ps_1 \gg ps_2 \xrightarrow{a} ps'_1 \gg ps_2}$
	$\dfrac{ps_1 \xrightarrow[\widehat{P}]{\tau} ps'_1}{ps_1 \gg ps_2 \xrightarrow[\widehat{P}]{\tau} ps'_1 \gg ps_2}$
(Parallel)	$\dfrac{ps_1 \xrightarrow{a} ps'_1 \quad a \in Act}{ps_1 \vert\vert\vert ps_2 \xrightarrow{a} ps'_1 \vert\vert\vert ps_2}$
	$\dfrac{ps_1 \xrightarrow[\widehat{P}]{\tau} ps'_1}{ps_1 \vert\vert\vert ps_2 \xrightarrow[\widehat{P}]{\tau} ps'_1 \vert\vert\vert ps_2}$
	$\dfrac{ps_1 \xrightarrow{a} ps'_1 \quad a \in Act}{ps_2 \vert\vert\vert ps_1 \xrightarrow{a} ps_2 \vert\vert\vert ps'_1}$
	$\dfrac{ps_1 \xrightarrow[\widehat{P}]{\tau} ps'_1}{ps_2 \vert\vert\vert ps_1 \xrightarrow[\widehat{P}]{\tau} ps_2 \vert\vert\vert ps'_1}$

Table 3.5. *Semantic rules of agent plan configurations*

The consistency of a set of intentions I can also be checked over some trace σ of the CPS, in particular in two extreme cases:

– if $|end(\sigma)| = |I|$, meaning that all the intentions of I are consistent;

– if $|end(\sigma)| = 0$, there is no satisfied intention, so the agent plan $\overline{P}$ is contextually inappropriate with respect to the set of intentions I.

3.6.3. *Application to the scenario*

Consider again the scenario of section 3.2. Alice and Bob's agents are specified separately but can pervasively coordinate to help achieving the intentions of their assisted users. Intention plans are composed from elementary plans which are viewed as concurrent processes, whose executions are terminated by an $exit$.

Alice's scenario	
q_0^A	$I_0 = \{meeting(Bob, \ell_1)\}$ $\boxed{P_0} = (meet(Bob); exit_m, \widehat{P_m})$
q_1^A	$I_1 = \{meeting(Bob, \ell_1), asking(Bob, get\_copies(\ell_2))\}$ $\boxed{P_1} = (meet(Bob); exit_m, \widehat{P_m}) \mid\mid\mid (Bob!(get\_copies(\ell_2)); exit_a, \widehat{P_a})$

Bob's scenario	
q_0^B	$I_0 = \{waiting(\nu), meeting(Alice, \ell_1)\}$ $\boxed{P_0} = (Alice?(\nu); exit_w, \widehat{P_w} \mid\mid\mid (move(\ell_1); meet(Alice); exit_m, \widehat{P_m})$
q_1^B	$I_1 = \{meeting(Alice, \ell_1), getting\_copies(\ell_2)\}$ $\boxed{P_1} = (get\_copies(\ell_2); confirm!(Alice); exit_g, \widehat{P_g}) \mid\mid\mid (move(\ell_1);$ $meet(Alice); exit_m, \widehat{P_m})$

Table 3.6. *Agent plans of Alice and Bob*

The mental process of an HoA agent can order its set of intentions, according to some preferences of the assisted user. For instance, the set of intentions of Bob's agent in q_1^B, $I_1 = \{meeting(Alice, \ell_1), getting\_copies(\ell_2)\}$, contains two concurrent intentions of the same weight such that $weight(meeting(Alice, \ell_1)) = weight\ (getting\_copies(\ell_2))$. The pairs $(E_m, \widehat{P_m})$ and $(E_g, \widehat{P_g})$ are two intention plan configurations corresponding to Bob. By using the $libp$, $options$ and $plan$ mappings, the built agent plan configuration is: $\boxed{P_1} ::= \Big((E_g, \widehat{P_g}) \mid\mid\mid (E_m, \widehat{P_m})\Big)$. The intention plan $\widehat{P_m}$ is used to satisfy the intention $meeting(Alice, \ell_1)$ and has only one elementary plan P_m expressed by $E_m ::= move(\ell_1); meet(Alice); exit_m$. The second intention plan $\widehat{P_g}$ corresponds to the intention $getting\_copies(\ell_2)$ and also has one elementary plan P_g expressed by $E_g ::= getc(\ell_2); confirm; exit_g$.

With respect to the HoA configuration q_1^B, the initial planning state is directly given from the agent plan configuration $[\overline{P_1}]$. One of possible derivations from this configuration consists in performing the $getc$ action. So, the configuration changes to $(confirm; exit_g, \widehat{P_g}) \mid\mid\mid (move(\ell_1); meet(Alice); exit_m, \widehat{P_m})$.

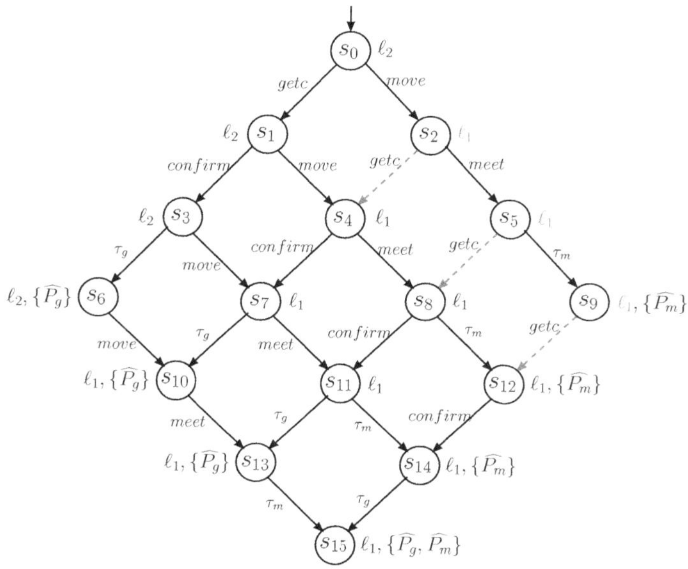

Figure 3.6. *The CPS_B corresponding to the plan $\overline{P_1}$ of q_1^B. For a color version of this figure, see www.iste.co.uk/elfallah/smartcities.zip*

The CPS of Bob's agent, denoted by CPS_B, is illustrated in Figure 3.6. It is built from the initial CPS state, $s_0 = ([\overline{P_1}], \ell_2, \emptyset)$, taking into account the current location ℓ_2 of Bob.

In the figure, the dashed edges represent the unrealized transitions from the states $s \in \{s_2, s_5, s_9\}$, because $pre(getc) = \ell_2 \notin \mathcal{L}\ (s)$. An example of maximum trace derived from s_0 is the following, expressing that Bob got the copies before moving to the meeting with Alice:

$$\Big((E_g, \widehat{P_g}) \mid\mid\mid (E_m, \widehat{P_m}), \ell_2, \emptyset\Big) \xrightarrow{getc} \Big((E'_g, \widehat{P_g}) \mid\mid\mid (E_m, \widehat{P_m}), \ell_2, \emptyset\Big) \xrightarrow{confirm}$$

$$\Big((E''_g, \widehat{P_g}) \mid\mid\mid (E_m, \widehat{P_m}), \ell_2, \emptyset\Big) \xrightarrow[\widehat{P_g}]{\tau} \Big((E_m, \widehat{P_m}), \ell_2, \{\widehat{P_g}\}\Big) \xrightarrow{move(\ell_1)}$$

$$\Big((E'_m, \widehat{P_m}), \ell_1, \{\widehat{P_g}\}\Big) \xrightarrow{meet} \Big((E''_m, \widehat{P_m}), \ell_1, \{\widehat{P_g}\}\Big) \xrightarrow[\widehat{P_m}]{\tau}$$

$$\Big((stop, \widehat{P_m}), \ell_1, \{\widehat{P_g}, \widehat{P_m}\}\Big)$$

In this short example, there are ten maximum traces. It is worth noting that the CPS could grow exponentially depending on the current intentions of the agent, hence on the related actions in plans. Nevertheless, we consider that the number of intentions to be dealt dynamically is reasonably small; moreover, the CPS structure is only partial since actions are only allowed with respect to the current spatio-temporal context of the agent[1].

3.7. Spatio-temporal guidance from past experiences

The proposed CPS-based guidance offers a forward view, assuming the correct performances of actions; however, depending on the context when definitively executing these actions, some failure could appear. In this section, we cope with this issue by taking profit from the past experiences of executions. By the use of an augmented architecture and the knowledge of some agent working preferences, we develop an enriched CPS structure to choose among the possible maximum paths. Hence, this yields a way to efficiently reinforce the guidance mechanism.

3.7.1. *Contextual planning architecture*

Figure 3.7 brings out a functional view of our enriched agent architecture. We focus on the planning process which is developed from two sub-processes.

The Plan Factory process builds the so-called *agent plan* which globally corresponds to the current set of weighted intentions. As we know, the *LibP library* is used to exploit the different alternative of plans that must be specified for each intention.

The Reinforced Guidance process is able to build the CPS from an agent plan and to enrich its information taking profit from the past-experiences of actions. This results in a structure called *CPS with Learning* or *CPS-L* for short, The information of past-experiences yielded by the execution process is stored and handled in a new module, called *Learning Contextual Experiences* or *LCE* for short.

The guidance offered by the refined planning process is driven by algorithm 3.1, yielding an *optimal maximum trace* σ_{opt}. As the planning process acts as a service provider, it must also notify the case where there is no possible maximum trace. Actually, this is useful to let the mental process revise its planning process request in accordance. Pay attention that there are now three input parameters to drive the

1 Intuitively speaking, the achievements of intentions are more or less costly depending on the plans they involve.

planning process. The set I of weighted intentions is augmented by a strategy $\mathcal{S}$ and a balance $\mathcal{B}$, which are detailed in the next sections.

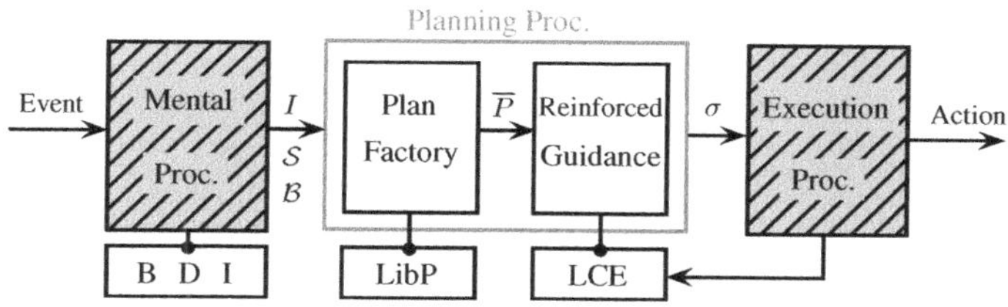

Figure 3.7. *Planning process architecture. For a color version of this figure, see www.iste.co.uk/elfallah/smartcities.zip*

Algorithm 3.1. Spatio-temporal planning process

Require:
I: set of weighted intentions;
$\mathcal{S}$: relevance strategy;
$\mathcal{B}$: balance proportions;
LCE: Learned Contextual Experiences module;

Build $\overline{P}$ from I;
Construct the CPS from $\overline{P}$;
Extract Σ_{MAX} from the CPS;
if $\Sigma_{MAX} \neq \emptyset$ **then**
    Enrich the *CPS-L* from the CPS and LCE;
    Order Σ_{MAX} from the *CPS-L* and $\mathcal{B}$;
    Return σ_{opt} among the ones of Σ_{MAX};
else
    Return $\emptyset$;
end if

3.7.2. *Learning actions from past experiences*

We demonstrate first how the planning process, in relation to the execution process, can quantify the performance of situated actions according to a current location and a current execution time. A preliminary stage of this quantification consists of extracting pertinent past-experiences from the LCE module.

3.7.2.1. *Data acquisition*

The performance of an action a is evaluated with respect to a given context; however, we focus on location. Each concrete performance of a in some location is considered as an *action experience*.

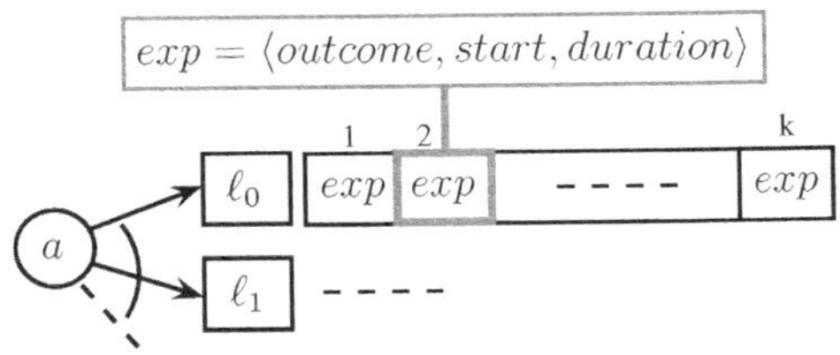

Figure 3.8. *Learned Contextual Experiences (LCE) of an action a. For a color version of this figure, see www.iste.co.uk/elfallah/smartcities.zip*

DEFINITION 3.5 (Action experience).– *An experience exp related to the performance of an action a in a location ℓ is a tuple $\langle outcome, start, duration \rangle$, where:*

– *$outcome \in \{-1, 1\}$, is the result of the run performing a, respectively a failure or a success;*

– *$start \in \mathbb{R}^+$, is the start date for the run of a;*

– *$duration \in \mathbb{R}^+$, is the action duration which provides that a was successfully terminated, undefined otherwise.*

The structure represented in Figure 3.8 is generically called the *learned contextual experiences* (*LCE*). Regarding the action a, LCE_a shows different FIFO queues of past-experiences of a, distinguishing the different locations where the action was performed. More precisely, if an action a is performed in a location ℓ, the agent may push a new experience, say exp, such that $LCE_a(\ell) = LCE_a(\ell) \cup \{exp\}$. Observe that the elements of the queue $LCE_a(\ell)$ are ordered on their $start$ dates, so that the last recorded past experience is at the top of the queue. Moreover, k represents the effective size of $LCE_a(\ell)$. Regarding any experience exp of a queue $LCE_a(\ell)$, we denote by $index(exp)$ the position of exp in the queue. Furthermore, the three components of a past-experience exp are respectively denoted by $exp.outcome$, $exp.start$ and $exp.duration$.

3.7.2.2. *Data relevance strategies*

The strategy information specified by the agent is given by the following definition and described below.

DEFINITION 3.6 (Agent strategy).– *The queues of* LCE *are parameterized by the agent strategy* $\mathcal{S} = \langle K, forget, M, \mathcal{C}\rangle$ *where:*

– K *is the maximum size of the queue;*

– $forget : 1..K \to \mathbb{R}^+$ *is the forgetting function, yielding a relevance weight for each experience;*

– $M \leq K$ *is the maximum number of filtered experiences;*

– $\mathcal{C}$ *is a periodic classification, e.g. daily, weekly, monthly or annual, applied in a modulo operation over the start dates of the queue.*

Defined over any queue, $filter$ *is a time filtering function yielding a sub-queue according to* M *and* $\mathcal{C}$*.*

$index$	1		5	6		10	11	12		18	k
$LCE_{getc}(\ell_2)$		...	A	B	...	C	D	E	...	F	...

$filter_{6,daily}(10)$

Experiences	$outcome$	$mod_{daily}(start)$	$duration$	$index$	$forget(index)$
A	-1	8.29	-	5	0.20
B	1	9.50	3.00	6	0.16
C	1	9.05	2.10	10	0.10
D	1	10.78	3.40	11	0.09
E	-1	10.05	-	12	0.08
F	1	11.05	5.12	18	0.05

Table 3.7. *LCE of the action* $getc$ *in the location* ℓ_2

3.7.2.2.1. The forgetting strategy

For all the queues in LCE_a, the forgetting function associates a *relevance weight* for each experience stored in the queue. As an interesting case illustrated in Table 3.7, the forgetting function $forget(index) = \frac{1}{index}$ is used, yielding much more relevance for any experience in the queue than another one of greater index.

For instance, the experience C stored at the index 10 in the queue $LCE_{getc}(\ell_2)$ has a forget value of $forget(10) = 0.10$, whereas the experience F stored at the index 18 has a forget value of $forget(18) = 0.05$.

Observe that every queue $LCE_a(\ell)$ is bounded by K positions. This allows us to tackle the data explosion problem implied by the consideration of many experiences within the LCE module. In the case where the queue is full, the addition of a new experience causes the removal of the oldest one.

3.7.2.2.2. The time filtering strategy

To operate the selection in some queue, say $LCE_a(\ell)$, both the start dates of the experiences and the *current date value* '$date$' are evaluated through some periodic classification $\mathcal{C}$. For this purpose, we introduce the function $mod : \mathcal{C} \times \mathbb{R}^+ \longrightarrow \mathbb{R}^+$, such that $mod_\mathcal{C}(date)$ corresponds to the start date modulo the classification period $\mathcal{C}$ (we consider standardly that a date can be viewed either in its textual form or like a real timestamp value). For instance, if $date =$ `'Monday 10 February 2015, 10:00'`, then $t = mod_{daily}(date) = 10$, whereas $mod_{weekly}(date) =$ `'Monday 10'`. We filter the experiences in the queue to only consider the ones having the smallest time intervals with respect to t, i.e. $|t - mod_\mathcal{C}(exp.start)|$. The mapping $filter_{M,\mathcal{C}}(t)$ of $LCE_a(\ell)$ specifies that M experiences must be selected. In case M is greater than the size k of the queue, all the past-experiences of the queue are considered.

In Table 3.7, the applied filtering is $filter_{6,daily}(10) = \{A, B, C, D, E, F\}$ which means that $LCE_{getc}(\ell_2)$ is filtered on the six closest experiences, with respect to $t = mod_{daily}(date) = 10$. As illustrated in Figure 3.9 for $mod_{daily}(t)$ over the start dates of $LCE_{getc}(\ell_2)$, the modulo operation applied to the queue graphically yields a spiral ribbon, the rings of which correspond to the successive periods, e.g. days in our example. From a daily perspective, the start dates of the experiences A, B and C occur before t and the ones of D, E and F occur after t.

In this example, we have taken the following instantiated strategy: $\mathcal{S}_1 = (20, \frac{1}{index}, 6, daily)$.

3.7.2.3. *Computing the expected performance and expected duration for an action*

For each non-empty queue $LCE_a(\ell)$, the expected performance $EP_a(\ell)$ represents the performance of a in some location ℓ. Its computation is based on M experiences, filtered from $LCE_a(\ell)$, such that

$$EP_a(\ell) = \frac{\sum_{exp}^{filter_{M,\mathcal{C}}(t)} exp.outcome * forget(index(exp))}{\sum_{exp}^{filter_{M,\mathcal{C}}(t)} forget(index(exp))} \qquad [3.1]$$

The expected performance value $EP_a(\ell)$ belongs to $[-1, 1]$. It is the weighted means, whose coefficients are given by the forgetting function in respect to the indices of considered experiences in $LCE_a(\ell)$. Hence, when performing a in ℓ, the closer '1' is to $EP_a(\ell)$, the greater the chance of success, whereas the closer to '-1', the greater the risk of failure.

In the case of $LCE_a(\ell) = \emptyset$, meaning that the running of a in ℓ has not been already explored, we choose $EP_a(\ell) = 0$ to privilege the exploration against every (bad) case such that $EP_a(\ell) < 0$.

The computation of the expected duration of the action a in ℓ, denoted by $ED_a(\ell) \in \mathbb{R}^+$, is similar to the expected performance one. It corresponds to the weighted means of the durations of M filtered experiences. Again, the weights correspond to forget function:

$$ED_a(\ell) = \frac{\sum_{exp}^{filter_{M,c}(t)} exp.duration * forget(index(exp))}{\sum_{exp}^{filter_{M,c}(t)} forget(index(exp))} \quad [3.2]$$

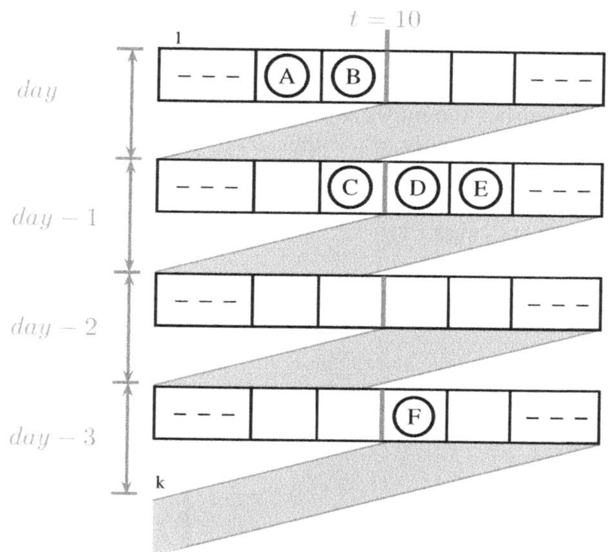

Figure 3.9. *The example of the queue $LCE_{getc}(\ell_2)$. For a color version of this figure, see www.iste.co.uk/elfallah/smartcities.zip*

Coming back to the frame example, from $t = 10$, we obtain $EP_{getc}(\ell_2) = 0.19$. Moreover, $ED_{getc}(\ell_2) = 3.16$, which in real time, stands for 3h10mn. Observe that the past experience F has a weak impact on $ED_{getc}(\ell_2)$ despite its important duration (5.12). In fact, the forgetting function applied to its (important) 18 index makes it negligible compared to the other filtered past experiences of lower indices.

3.7.3. *Spatio-temporal guidance*

3.7.3.1. *Contextual planning system with Learning (CPS-L)*

The *CPS-L structure* focuses on the maximum traces Σ_{MAX} of the CPS, augmented by the different values $EP_a(\ell)$ and $ED_a(\ell)$, for each mentioned action a to be performed in some location ℓ.

DEFINITION 3.7 (Contextual Planning System with learning).– *With respect to the set of intentions, the CPS-L is a tuple* $\langle CPS, \mathcal{EP}, \mathcal{ED}\rangle$ *where:*

– $CPS = \langle S, s_0, Tr, \mathcal{L}, \mathcal{T}\rangle$ *s.t.* $\Sigma_{MAX} \subseteq 2^{Tr}$ *is the set of the maximum traces of the CPS;*

– $\mathcal{EP}$ *is a mapping from* Tr *to* $[-1, 1]$, *s.t. from each transition* $tr = (s, a, s') \in \Sigma_{MAX}$, $\mathcal{EP}(tr) = EP_a(\mathcal{L}(s))$;

– $\mathcal{ED}$ *is a mapping from* Tr *to* $\mathbb{R}^+$, *s.t. from each transition* $tr = (s, a, s') \in \Sigma_{MAX}$, $\mathcal{ED}(tr) = ED_a(\mathcal{L}(s))$.

Figure 3.10 is a representation of the CPS-L structure derived from the CPS of Figure 3.6. It distinguishes the maximum traces (with possibly common prefix), and over two of them, we explicitly show the labels attached to the transitions. For instance, the values $EP_{getc} = 0.19$ and $ED_{getc} = 3.16$ are attached to the transition $(s_0, getc, s_1)$. These values are computed from the experiments selected from $LCE_{getc}(\ell_2)$, specified in Table 3.7.

From any CPS-L, it is straightforward to extend the *expected qualities* of the used situated actions, to every maximum trace $\sigma \in \Sigma_{MAX}$, to compare them, as follows:

$$QP(\sigma) = \frac{\sum\limits_{tr \in Tr}^{\sigma} \mathcal{EP}(tr)}{|\sigma|} \quad [3.3]$$

$$QD(\sigma) = \sum_{tr \in Tr}^{\sigma} \mathcal{ED}(tr) \quad [3.4]$$

To compare the traces in-between them, we normalize all expected quality values to be in $[-1, 1]$, such that '-1' represents the worst case and '1' the best one. This is already done for QP, moreover to normalize QD, we consider the extreme values $QD_{min} = min\,(QD(\sigma) \mid \sigma \in \Sigma_{MAX})$ and $QD_{max} = max\,(QD(\sigma) \mid \sigma \in \Sigma_{MAX})$; hence, the *normalized quality of duration* is featured by the $NQD(\sigma)$ expression, for any trace σ.

$$NQD(\sigma) = 1 - \left(\frac{QD(\sigma) - QD_{min}}{QD_{max} - QD_{min}} * 2\right) \quad [3.5]$$

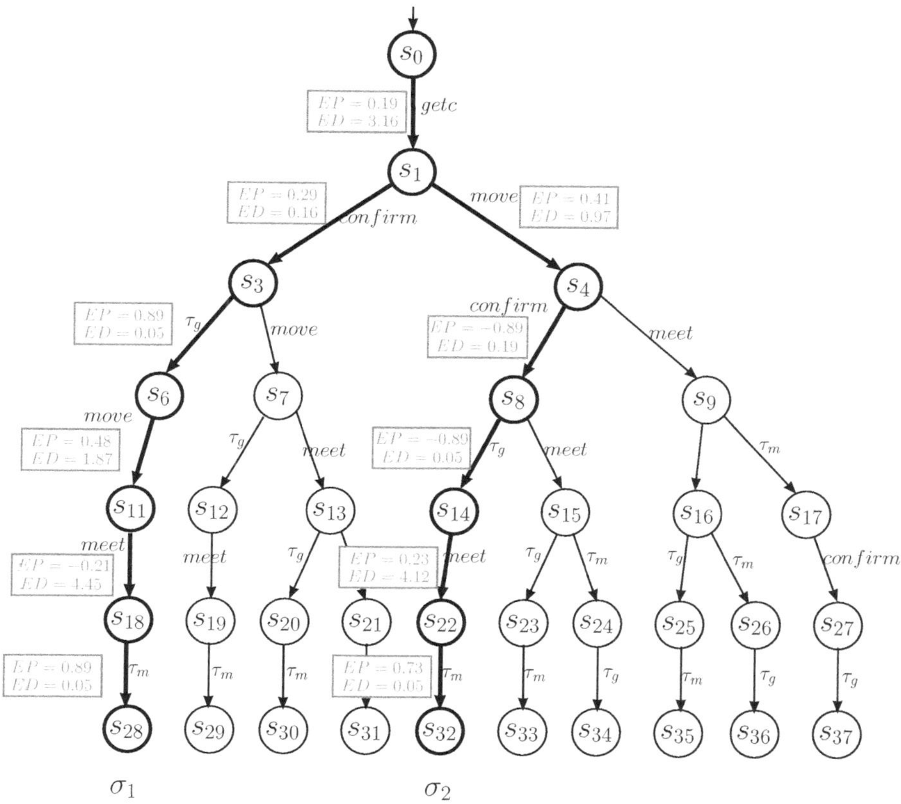

Figure 3.10. *Maximum traces and $CPS\text{-}L_B$ for the plan $\overline{P_1}$ of q_1^B. For a color version of this figure, see www.iste.co.uk/elfallah/smartcities.zip*

As shown in Figure 3.11, the traces, the quality of which enters the right-upper part of the figure, are the best ones, and the ones entering the left-lower part are the worst ones. The other traces having a single high value either for QP or NQD enter the two other parts of the figure.

3.7.3.2. *Optimal trace of the CPS-L*

To select an optimal trace, we order the set Σ_{MAX} of maximum traces regarding the respective values, QP and NQD, considered separately. This allows the planning process to select an optimal trace, either with the lowest duration or with the maximum performance (i.e. with the minimum risk of failure).

As a compromise between performances and durations, we also assume that the mental process can specify a preference *balance* to the planning process. This balance denoted by $\mathcal{B}$ is expressed as a proportion of the QP and NQD values, to obtain a *global quality* value $Q_{\mathcal{B}}(\sigma)$ belonging to $[-1, 1]$ for each of the maximum traces ($\sigma \in \Sigma_{MAX}$). In this case, the optimal trace σ_{opt} corresponds to the maximum trace having the best global quality value.

$$Q_{\mathcal{B}}(\sigma) = \mathcal{B}_P * QP(\sigma) + \mathcal{B}_D * NQD(\sigma) \quad with \quad \mathcal{B}_P + \mathcal{B}_D = 1 \qquad [3.6]$$

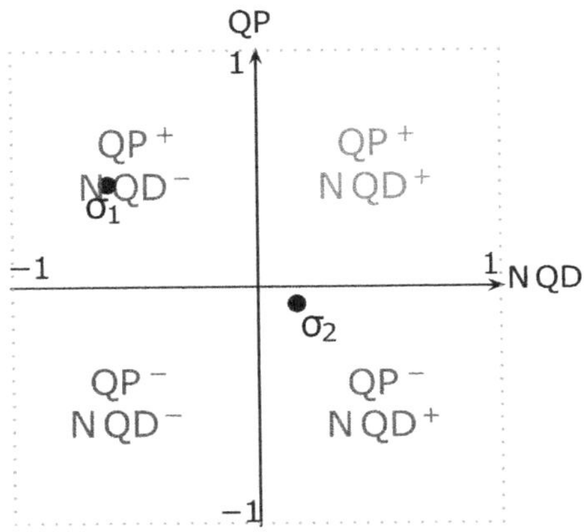

Figure 3.11. *Balance between the $QP(\sigma)$ and the $NQD(\sigma)$ of a trace σ. For a color version of this figure, see www.iste.co.uk/elfallah/smartcities.zip*

Table 3.8 highlights the global qualities of two maximum traces σ_1 and σ_2 of Figure 3.10, knowing that over the whole set of maximum traces, we obtain $QD_{min} = 7.58$ and $QD_{max} = 10.23$. Assuming the preference balance $\mathcal{B}1$ such that $\mathcal{B}1_P = 0.30$ and $\mathcal{B}1_D = 0.70$, it appears that $Q_{\mathcal{B}1}(\sigma_2)$ is greater than $Q_{\mathcal{B}1}(\sigma_1)$, so the maximum trace σ_2 can be offered to the execution process as the optimal maximum trace. In contrast to $\mathcal{B}2$ ($\mathcal{B}2_P = 0.70$, $\mathcal{B}2_D = 0.30$), σ_2 is the one that can be offered ($Q_{\mathcal{B}2}(\sigma_1) > Q_{\mathcal{B}2}(\sigma_0)$).

Σ_{MAX}	QP	QD	NQD	$Q_{\mathcal{B}1}$	$Q_{\mathcal{B}2}$
σ_1	0.42	9.74	−0.63	−0.31	0.10
σ_2	−0.04	8.54	0.27	0.17	0.05

Table 3.8. *Computed quality values of σ_1 and σ_2 maximum traces*

3.7.3.3. *Exploration versus exploitation*

The balance technique presented in the previous section must be adapted to tackle the cases where the quality of duration (QD) cannot be computed for a trace:

– a trace containing $EP = 0$ labeled transition may correspond to an untested action;

– a trace containing $EP{=}{-}1$ labeled transition corresponds to a failed action.

A rough approach would consist of abstracting these cases when searching the balance selection. However, enabling $EP = 0$ labeled transition makes exploration cases possible. Moreover, even an action declared to have failed could perform correctly, since our approach based on past-experiences is only heuristics.

In fact, our approach depends on the mental attitudes of the agent:

– a *conservative agent* only accepts the nominal cases. For such agents, all the transitions having an undefined ED value must be discarded as well as the related traces. The maximum trace selection typically exploits the QP and QD information of the remaining maximum traces, and the selection is finally driven by the proposed balance approach.

– a *curious agent* rather gives a chance to play any $EP{=}0$ or $EP{=}{-}1$ transition (with QD undefined). Due to the presence of undefined QD values, the maximum trace selection is performed over the QP information only.

3.8. Conclusion

The HoA model formally represents the behavior of an AmI agent able to reason, communicate and move. Agent dynamicity and context-awareness are handled due to the fact that both BDI and planning states are changed consistently and adequately to the perceptions of new events.

The proposed AgLOTOS is an algebraic language specifically defined for our original planning process, to be embedded within the agent software architecture. It appears to be a powerful and compact way to formally express an agent plan as a set of concurrent and partially ordered processes. The presented scenario has shown how the AgLOTOS language is expressive due to modularity concepts and concurrency operators.

Much more than a simple plan to satisfy an intention, an agent plan is the expression of all the possible execution plans that can be built with respect to some set of intentions. Knowing in addition that the updates of intentions can be handled on the fly at the AgLOTOS level, as demonstrated in [BOU 15]. An AmI agent can be viewed as having only one agent plan, consistent with the whole set of intentions and updated all along the evolution of the agent behavior.

The planning process basically comes from its underlying CPS structure, working on the causality of situated actions and the terminations of intention plans within the agent plan. This offers an interesting way to guide the agent contextually. Basically, the guidance service concentrates on the search of a maximum execution plan, satisfying the whole set of intentions. In addition, the CPS-L structure yields more refined results qualifying the different maximum executions plans by exploiting the past-experiences of situated actions it promotes. Actually, we show that this can be used to quantify the situated actions, up to obtaining in general, performance and duration values for each maximum trace. We also discussed how the guidance service can balance between the performance and duration criteria in accordance with the specification of the AmI agent.

Moreover, as an interesting extension to cope with an unexpected context, our approach integrates the possibility of exploration regarding some situated actions, in two cases overpassing the simple knowledge given by the related past-experience: when some situated action does not refer to any pertinent past-experiences or when it only refers to failure. Our planning process can, thus, adapt to the conservative and curiosity attitudes of the agent, when returning an optimal maximum plan to it. It can work in initial or degraded cases.

As with any learning approach, the pertinence of the past-experiences to consider is a cornerstone. In our approach, the past-experiences which are not in strong relation with the situated actions of the CPS-L structure are filtered. A periodic classification of past experiences can be exploited for that, moreover, the forgetting strategy allows us to privilege the impact of the more recent past-experiences among the filtered ones.

3.9. Bibliography

[AIR 08] AIRIAU S., PADGHAM L., SARDINA S. *et al.*, "Incorporating Learning in BDI Agents", *Proceedings of ALAMAS - ALAg*, 2008.

[BOU 15] BOUKHARROU R., CHAOUCHE A.-C., EL FALLAH SEGHROUCHNI A. *et al.*, "Dealing with temporal failure in ambient systems: a dynamic revision of plans", *Journal of Ambient Intelligence and Humanized Computing*, vol. 6, no. 3, pp. 325–336, Springer Berlin Heidelberg, 2015.

[BRI 88] BRINKSMA E. (ed.), ISO 8807, LOTOS – A Formal Description Technique Based on the Temporal Ordering of Observational Behaviour, standard 1988.

[CAL 14] CALVARESI D., CLAUDI A., DRAGONI A.F. *et al.*, "A goal-oriented requirements engineering approach for the ambient assisted living domain", *PETRA'14*, New York, USA, ACM, pp. 20:1–20:4, 2014.

[CHA 14] CHAOUCHE A.-C., EL FALLAH SEGHROUCHNI A., ILIÉ J.-M. *et al.*, "A higher-order agent model with contextual management for ambient systems", *Transactions on Computational Collective Intelligence XVI*, vol. 8780 of *Lecture Notes in Computer Science*, pp. 146–169, Springer, Berlin Heidelberg, 2014.

[GEO 98] GEORGEFF M.P., PELL B., POLLACK M.E. *et al.*, "The belief-desire-intention model of agency.", in MÜLLER J.P., SINGH M.P., RAO A.S., (eds), *ATAL*, vol. 1555 of *Lecture Notes in Computer Science*, Springer, pp. 1–10, 1998.

[GUI 12] GUIVARCH V., CAMPS V., PÉNINOU A., "Context awareness in ambient systems by an adaptive multi-agent approach", in PATERNÒ F., RUYTER B., MARKOPOULOS P. (eds), *Ambient Intelligence*, vol. 7683 of *Lecture Notes in Computer Science*, Springer Berlin Heidelberg, pp. 129–144, 2012.

[MEN 07] MENEGUZZI F., ZORZO A.F., DA COSTA MÓRA M. *et al.*, "Incorporating planning into BDI agents", *Scalable Computing: Practice and Experience*, vol. 8, pp. 15–28, 2007.

[NUN 14] NUNES I., LUCK M., "Softgoal-based plan selection in model-driven BDI agents", *AAMAS'14*, pp. 749–756, 2014.

[OLA 13] OLARU A., FLOREA A.M., EL FALLAH SEGHROUCHNI A., "A context-aware multi-agent system as a middleware for ambient intelligence", *MONET*, vol. 18, no. 3, pp. 429–443, 2013.

[PRE 12] PREUVENEERS D., NOVAIS P., "A survey of software engineering best practices for the development of smart applications in ambient intelligence", *JAISE*, vol. 4, no. 3, pp. 149–162, 2012.

[RAO 95] RAO A.S., GEORGEFF M., "BDI Agents: from theory to practice", *First International Conference on Multi-Agent Systems (ICMAS-95)*, S. Francisco, CA, pp. 312–319, June 1995.

[SAR 06] SARDINA S., DE SILVA L., PADGHAM L., "Hierarchical planning in BDI agent programming languages: a formal approach", *AAMAS'06*, pp. 1001–1008, 2006.

4

A Multi-Agent Middleware for Deployment of Ambient Applications

Ambient intelligence (AmI) provides a vision of the information society where heterogeneous hardware entities are disseminated in the environment and used by intelligent agents to provide ubiquitous applications. However, this horizontal integration has not been reached yet. This is due to the heterogeneity and dynamicity of such systems and also the lack of mechanisms to reason on it. In this chapter, we present an overview of the actual use of connected devices and ambient applications. We then propose a model that handles heterogeneity by describing both the needs of the applications and the properties of the available hardware infrastructure. This description allows us to present mechanisms and algorithms to find in the hardware infrastructure, the available entities that can support the deployment and the running of distributed applications. Finally, we present a middleware, based on a multi-agent system (MAS), for the deployment and the configuration of distributed applications. This middleware deals with the dynamicity of smart environments by promoting context-aware local reasoning, ensuring privacy and scalability.

4.1. Introduction

Some research domains like ambient intelligence or Internet of Things appeared in the early 2000s with technological improvements and ongoing miniaturization of electronic entities. These electronic and information devices can communicate and exchange data more and more easily. This leads to new application fields like smart cities, characterized by a high heterogeneity. The hardware is always more powerful: we have to deal with computers, but also with smartphones, micro-computers like TV-boxes or other sensors and connected devices. The communication means and

---

Chapter written by Ferdinand PIETTE, Amal EL FALLAH SEGHROUCHNI, Patrick TAILLIBERT, Costin CAVAL and Cédric DINONT.

protocols are application specific: some applications use Wifi as they need huge bandwidth over short range and others use SigFox or LoRa to communicate very few data over a very long range. The speed, the amount of data and the semantics of data streams can also be very flexible and tend to increase so as to render services and applications more useful: energy consumption data can be acquired once a month for a billing service, tenth per hour for a heating application and every second for activity monitoring.

With the increase in heterogeneity and amount of data, we need systems that ease the development and the deployment of applications. These applications have to handle and process large amounts of data in real time. Some techniques exist in the field of Big Data to process this voluminous data. However, they are reserved for processing in data centers regardless of the data source. In this kind of configuration, the heterogeneity of hardware is small and generally neglected. These techniques are well adapted to process data generated by an ambient system; they cannot be used in them. Moreover, Big Data techniques are very beneficial but can lead to a Big Brother effect due to the free flow of the data in the cloud. To prevent this effect while allowing the use of data on a large scale, we need to deal with data privacy, favoring a local processing with an agreed data sharing instead of a cloud processing.

We propose a multi-agent middleware to ease the development, deployment, configuration and monitoring of applications for ambient systems.

The deployment of applications on an ambient platform should be, at least, as easy as the downloading of applications on smartphones from a store. To reach an easy deployment, it is appropriate to decorrelate the hardware and the software parts. Our middleware decorrelates applications from hardware infrastructures separately by describing with different levels the various entities of the system. The specifications and properties of the available hardware entities of the infrastructure and the requirements of ambient applications are modeled with graphs. These descriptions allow us to reason about the deployment of applications on a heterogeneous hardware infrastructure by using a graph-matching algorithm. This algorithm finds a graph homomorphism between the application graph and the hardware infrastructure graph: for each node of the application graph, the algorithm finds a hardware entity that matches the requirements of the application node. The solution of the deployment represents a subset of the hardware infrastructure on which the application will run.

The agent organization ensures data and resources privacy. Infrastructure agents are located geographically and manage part of the hardware infrastructure. Application agents manage the applications used on it and the data generated by these applications. An application is locally used and configured by infrastructure agents, using the graph-homomorphism algorithm, and if there is no solution, the agent will collaborate with others to use applications on the global hardware infrastructure of which no global and centralized representation exists. The spreading of data and the accessibility of resources are managed locally by these agents according to the description of the hardware and software infrastructure and the user preferences.

Finally, the mobility of the users, the applications and the data are taken into account, thanks to artifacts, generating contextual information and user agents, following the user, whose purpose is to adapt the deployment of applications depending on the context by coordinating the nearby infrastructure agents.

4.2. Challenges for ambient intelligence and Internet of Things

4.2.1. *Toward the heterogeneity of hardware and protocols*

4.2.1.1. *Ambient intelligence: a vision of the future*

Ambient intelligence (AmI) was born in the early 2000s with the technological improvements and ongoing miniaturization of electronic entities. Following on from pervasive computing [WEI 93], the IST advising group defines AmI as a vision of the information society where people are surrounded by intelligent intuitive interfaces that are embedded in all kinds of objects and an environment that is capable of recognizing and responding to the presence of different individuals in a seamless, unobtrusive and often invisible way [DUC 01]. Ambient applications have to deal with the user context and are characterized by a high dynamicity.

Lots of works in AmI research focus on the improvement of human interactions with intelligent applications. These improvements are made possible by the proposal of frameworks and platforms that facilitate the development of context-aware and dynamic applications. These platforms offer mechanisms to build such applications by handling data and events [HEL 04, JOH 02] or by wrapping hardware/software capabilities in agents [CHE 04, ELF 10]. Multi-agent systems are often used as they provide good properties and allow us to deal with autonomy, context management, robustness, fault tolerance and scaling. However, these works assume that an underlying ubiquitous infrastructure already exists [O'H 12].

Other platforms and middleware for AmI were developed to allow the development of applications without being restrictive in terms of underlying software and hardware platforms, neither in terms of assumed resources [GEO 10]. The middleware are often service-oriented architectures that aim to allow the integration of heterogeneous technologies [STA 13]. These service-oriented architectures use available resources present in the system. The connected devices and hosted applications are abstracted into services, which are composed or orchestrated by the middleware, according to the user needs. All these resources are already used and efficient. No reasoning is made on the location of these resources: e.g. an image-processing service will be hosted on a specific hardware entity. The user will use this particular service to make a processing on his image regardless of the bandwidth constraint between the service hardware entity and the client.

At last, very few works consider the deployment of applications. Braubach *et al.* [BRA 05a] proposed the deployment of a reference model based on MAS architecture (e.g. agent services) for using MAS applications. As an agent is a software entity, the deployment of agents does not have to deal with the high heterogeneity of hardware entities. Some other works in the service-oriented architectures (SOA) community use model-driven [EDW 04] or pattern-matching [ARN 08] to use and configure services. This work reasons on deployment patterns, which specify the structure and constraints of composite solutions on the infrastructure, to compose services. Contrary to our approach, the cited paper refers not to the localization of resources and installation of software, but rather to the binding of existing resources to provide the desired composition of services. This is realized using a centralized graph-matching algorithm that takes into account the various requirements for the given service. Flissi *et al.* [FLI 08] propose a meta-model for abstracting the concepts of the deployment of software over a grid. All these works have shortcomings when considering their use for using AmI applications on a hardware infrastructure. Some do not take into consideration the heterogeneity of the hardware and software, as well as the interaction between the two layers (i.e. software and hardware). Others do not tackle the privacy problem while some propose centralized solutions that are not scalable for real-life AmI applications.

4.2.1.2. *Internet of Things: an economic reality*

AmI focuses on the improvement of human interactions with intelligent applications, assuming that an underlying ubiquitous infrastructure already exists. In the meantime, Internet of Things (IoT) aims to provide this global infrastructure for the information society, enabling advanced services by interconnecting (physical and virtual) things based on existing and evolving interoperable information and communication technologies [ITU 12]. Main challenge of the IoT is to achieve full interoperability of interconnected devices while guaranteeing the trust, privacy and security of communications [ATZ 10].

Among the other technological improvements, we can discuss smartphones. In the 1990s, mobile phones were based on a closed model. Each manufacturer developed its own firmware and applications for the mobile phones they sold. After the 2000s, the first open firmware appeared (iOS, Android etc.). This firmware allows developers to design applications using the phone devices (GPS, camera, accelerometer etc.) without worrying about the phone model. The application deployment is also very simplified: a click from a store and here it is.

In line with this innovation, cloud computing allows access to data from anywhere with any device. This technology was used very quickly by developers for mobile phone applications. Thereby, the smartphone became the first widely available connected device that is able to synchronize its data with the Internet.

Currently, the IoT allows all kinds of electronic devices to generate and synchronize data. This leads to new perspectives and allows the user to imagine new kinds of applications using these data.

Multiple architectures and communication protocols make the design of applications a baffling problem, but it is also an advantage to produce high-performance systems and applications. However, there is still no work proposing mechanisms to take this high heterogeneity into account. Actually, we do have the intelligent applications and the hardware infrastructure, but nothing for reasoning about the deployment of these applications on the hardware infrastructure, choosing the good hardware entities to support the applications, according to the applicative constraints.

4.2.2. *Data transport and processing*

4.2.2.1. *The actual use of connected objects*

According to IDATE, Gartner or Cisco, there will be more than 50 billion connected devices in 2020 against only 4 billion in 2010 [EVA 11]. This will be the next economic revolution.

Most actual connected devices are capable of producing data and sending them via the Internet to a centralized server or cluster, often owned by the firm that commercializes the object. Processing the data is made on these servers to provide specific services or applications to the user. This is a vertical communication between connected devices and external cluster. The objects are not able to communicate and exchange data horizontally with each other. We can notice similarities with the mobile phone world in the 1990s. Though it is rare to find open objects that allow us to collect data directly without passing through a centralized server. This is due to the high heterogeneity of the hardware and protocols and the lack of mechanisms able to take it into account.

To connect all these devices to the Internet, we traditionally use gateways that interconnect different networks operating with different protocols. These gateways ensure transparency with the used protocols. However, by this abstraction, no reasoning is made on the different protocols specificities. Currently, no mechanism exists capable of reasoning on the hardware and protocol constraints. We can interconnect heterogeneous networks but without integrating these specificities.

4.2.2.2. *The growth in data exchanges*

In addition to the growth in the number of devices connected, the data exchanges will also increase. We notice societal changes in the use of data. Instead of getting the data when the application or the user needs it, the current trend is to store a huge amount of data and to think over how to use them to provide services.

This is, for instance, what is praised by the quantified-self movement. In this movement, people acquire personal data from their daily life and store them to monitor and improve their lifestyle. Initially, the data was collected by humans and stored on notebooks. But with the accession of connected devices, data are collected automatically, stored on remote servers, and data processing algorithms can now extract relevant information and automatically advises the users.

The actual trend is to real-time monitoring of all kinds of data. This leads to questions like network saturation or data-processing power.

4.2.2.3. *Evolution of data processing*

To handle the growth in data exchanges, processing techniques have evolved. This starts with the need for processing a large amount of data efficiently. This was widely studied in distributed computing. Currently, MapReduce [DEA 04] techniques are being used. It is a programing model developed in Google laboratories for processing large amount of data using a parallel and distributed algorithm on clusters. Each node of the cluster uses a Map function that applies a basic and sequential operation to a set of data. The output is composed of key/value data. They are redistributed to other nodes that apply a reduce step. This step processes the data by key and produce synthetic partial data that are collected to produce the final result.

Frameworks were built around this MapReduce technique. This is the case of the open source framework Hadoop, founded by Doug Cutting in 2009, developed and maintained by the Apache Foundation, used by lots of organizations like Facebook, Amazon or Yahoo. This framework helps us to design batch-oriented applications to process petabytes of data offline [WHI 12]. However, in addition to the increasing of the amount of data, it has to be processed in real time. That is why other frameworks based on MapReduce appear. Nathan Marz, for instance, developed Storm for Twitter's needs. It is an open source framework for processing data streams in real time over an homogeneous cluster with complex processing event model (CPE) [MAR 11].

The evolution of data processing is, of course, not centered on the MapReduce technique. Hadoop or Storm frameworks also provide a series of tools to manage data sharing, synchronization, storage and request. Tools like Zookeeper provide a centralized service for distributed synchronization or for maintaining the consistency of configuration information over a large distributed system. Distributed file systems like HDFS were also developed to provide transparency for the storage of files. To store data, databases have evolved too. From a relational row-oriented model, not only we passed on relational databases, column oriented, like Cassandra, HBase and now Google BigTable, used by Facebook, Twitter, Amazon or Google to efficiently deal with petabytes of data, but also there exist graph-oriented databases like OrientDb or Neo4j, used by eBay or Meetic. These kinds of NoSQL databases are especially efficient for storing changing data with no schemes. The graph representation allows us to make powerful graph queries, thanks to the associated query languages like Gremlin that exploit graph properties.

All these techniques, frameworks and tools have been designed to allow an efficient data processing over clusters of fairly homegeneous computers. This is done transparently for the user who designs his process, adds data as an input and gets the result as the output, regardless of the location of the processing or of the number of machines involved. Specificities of hardware components could be ignored in cluster environments but must be taken into account if one wants to use a distributed application in an ambient system.

4.2.3. *Management of data privacy*

4.2.3.1. *Privacy*

Privacy is defined by [WES 67] as "the claim of individuals, groups or institutions to determine for themselves when, how and to what extent information about them is communicated". Privacy in multi-agent systems has already been well-explored. Such *et al.* [SUC 14] categorize research on data privacy on different levels: collection, disclosure, processing and dissemination. Multi-agent system specificities have been used to propose different manners for handling data privacy. Some works focus on standards [BAR 06, KRU 10] and privacy policies [TEN 06, CRE 09, UDU 10], checked by agent brokers to control the disclosure of the data. Other works [RAM 04, SUC 12] use social relationships like trust, intimacy or reputation to select the agents with which the data can be shared. Trusted third parties are already used in [MEN 02, AIM 06, CIS 07] to make the data anonymous or the metadata (e.g. IP address, receiver or sender identity) and also to check disclosure authorizations. Last but not least, some works [ALB 10] focus on integrating secure communication in the agent platforms by using well-known encryption protocols. All these works use MAS to provide data privacy. In our work, as explained in section 4.4.2, we merely take advantage of MAS properties to handle the privacy of the hardware resources and of the structure of the system.

4.2.3.2. *Actual techniques: the bad things*

We have already seen that the development of the IoT allows the user to connect all kinds of devices, to generate, process and communicate huge amount of data over the Internet. These innovations and improvements lead to new application fields and innovations. However, the newness of these techniques induces lacks that prevent from good practices.

As we do not have easy ways to interconnect heterogeneous devices together, companies promote a vertical connection from the device to the company servers. Business models are conceived based on this architecture. This leads to closed devices. Data are directly acquired by the companies that own our data and use it to provide specific services to the users or to sell them to third parties. Anyway, the user has access to services and not to the data itself. This approach inevitably leads to a Big Brother effect: the user does not own his personal data.

Moreover, Cloud computing and Big Data techniques provide an easy way to synchronize, process and share data. These techniques, however, deliver data over the Internet and need a lot of bandwidth. That is sometimes essential, especially for social network applications, like Twitter, that have to process data on a global scale but this is not the case for most home or city applications. Monitoring our daily activities or managing a building produces data that can be handled and processed locally. In these cases, we do not need to export the data all across the world. What we need are multi-scale mechanisms and data privacy policies to export only some data, potentially aggregated, in the Cloud.

4.2.3.3. *Data privacy: societal issues*

We provide all these technologies to build smart applications for users. However, we need the users to accept these technologies and their applications. This will be possible if we are able to guarantee privacy of data.

Some commercial products have triggered controversies. That is the case for example of the XBox, one that constantly enables the microphone and reacts to some kind of sentences pronounced by the user. It is also the case for Google research mode on smartphones that reacts to the sentence "Ok Google Now". We have here two examples of potentially dangerous applications for data privacy: the system should acquire all the vocal data and send information to Microsoft or Google to detect potentially compatible sentences. We cannot know if these data are used for other purposes or not.

The question, however, is not to know if these systems exploit these data or not. It is interesting to note that the basic user cares about this life intrusion. Actually, nothing can reassure the user about his data privacy. Worse still, because of this lack of privacy mechanisms, people are increasingly reluctant to use Cloud technologies and services. Even if most people spread their personal data on Facebook or other applications without being concerned about their privacy and without realizing that their information is no longer owned by them, there are some, on the contrary, who militate against such abuses and whilst others succumb paranoia. To avoid this discomfort and promote ambient applications, we propose systems that guarantee data and resources privacy to the user.

4.3. Deployment of applications for ambient systems

4.3.1. *Reasoning about heterogeneity*

The aim of this research is to use the existing heterogeneous technologies and devices, developed separately by different developers, to be the support of intelligent applications. To achieve Smart Cities, we need an operational and interoperable telecommunication and energetic infrastructure. We also need some intelligent and context-aware applications. Both are already available but they lack the link between the available hardware infrastructure and the intelligent applications. We need some deployment mechanisms to automatically configure applications and use them on the existing hardware infrastructure. The hardware entities have to be used by several applications at the same time. We do not need to wait until we want a specific application, to set up the hardware entities, but on the contrary, we would like to use the existing infrastructure to use applications. We propose in this section a way to achieve this automatic deployment of applications.

We believe that the deployment of applications on an ambient platform should be at least as easy as the downloading of applications on smartphones from stores. To reach an easy deployment, it is appropriate to decorrelate the hardware and the software parts of an ambient platform. It is, however, more difficult in ambient systems than with smartphones because of the heterogeneity of the hardware and the software parts. Multiple architectures and communication protocols make the design of applications a baffling problem, but it is also an advantage to produce high-performance systems and applications. Some hardware architectures are designed to provide very low power consumption, but we can benefit from these different specificities by adapting architectures and protocols according to the need of each application.

We propose a model in which the hardware components are connected to support the running applications. These applications are a chain of functionalities provided by software entities. This chaining should be implemented beforehand and transparently by introducing projection algorithms for applications onto a heterogeneous hardware infrastructure. The specifications and properties of the available hardware entities of the infrastructure and the requirements of ambient applications are modeled with graphs. To project ambient applications onto the heterogeneous hardware infrastructure, we extend one of the classical graph-matching algorithms [MCK 14].

4.3.2. *Graph modeling*

The aim of our research is to decorrelate the hardware infrastructure from the applications running on it. To fulfill this purpose, we describe each entity of an AmI system and its relations with the other entities. This allows us to reason about the use of these entities by the applications: we will be able to choose, for each application, the best hardware entities to use to provide a good quality of service to the users. The application requirements and the available hardware entities are modeled with the same graph formalism.

4.3.2.1. *Hardware infrastructure graph*

The hardware infrastructure graph (or infrastructure graph) describes the available hardware entities, their properties and interactions. The first kind of node is the entity node. Two of these nodes can be linked by a relational node. The entity nodes represent a hardware entity available in the infrastructure, whereas the relational nodes inform about the semantic of the relation between the two entity nodes. This forms a bipartite graph: we cannot find a direct link between the two relational nodes or between two entity nodes. Each node of the graph, whatever its type, can be characterized by several properties. For instance, an available camera in the hardware infrastructure is represented by an entity node with a type property (its value is Camera), a frame rate property and an image size property.

The different property types allow us to model and describe the environment. These property types depend on the application field and the type of the entities.

Let us introduce an illustrative example of such a graph. Figure 4.1 is an example of a hardware infrastructure composed of two cameras and three computers linked by some communication channels. The bold rectangles represent entity nodes; the rounded rectangles represent relation nodes. Each entity has different properties and can communicate together. For instance, the IpCamera 1 has a configurable framerate: it can acquire 12, 24 or 48 images per second. The operating system of the computer 8 is based on Linux. The framerate, memory quantity and type of the operating system are properties of the infrastructure graph nodes.

4.3.2.2. *Application graph*

An application graph describes the structure and the requirements of an application needed to run it with good service quality. The functional structure part is an abstract view representing the different parts of the application interacting together. These different parts, called functionalities, can interact with the hardware infrastructure by using a hardware entity or using a software part on a hardware entity. The hardware requirement part forms a graph, using the same property set than the hardware infrastructure. This part informs us about the hardware entities and its properties required to use the application. To achieve a deployment, we would like to find a graph of the infrastructure graph that matches the hardware requirement sub-graph of the application graph.

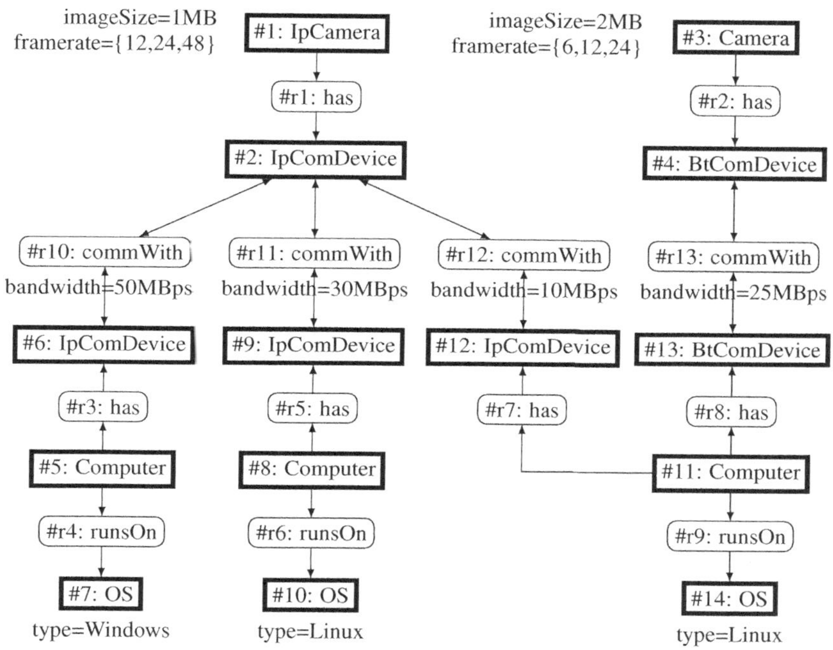

Figure 4.1. *Example of hardware infrastructure graph*

We can apply constraints on properties of the node of this hardware requirement sub-graph. These constraints allow us to automatically compute some property values in function of the available property values of the infrastructure graph. This allows the user to automatically configure the hardware entities and the application during the deployment process.

Consider that we would like to use a basic application that captures the video stream of a camera and processes it. This application is composed of two functionalities which exchange data. The first uses a camera available in the hardware infrastructure and the second should use a software entity on a computer for processing the camera datastream. Figure 4.2 shows the graph of this application. The bold rectangles represent entity nodes; the rounded rectangles represent relation nodes; and the diamonds represent functionalities. In this example, the application needs a camera with a minimum framerate of 10 images per second and the software entity that will execute the image processing to be used on a computer running on a

Linux operating system with at least 4 GB RAM. The camera used and the computer on which the software entity will be used have to communicate. The required bandwidth is a function of the image size of the camera and its framerate. This will allow us to automatically configure the framerate of the camera with the available bandwidth. In Figure 4.2, we see the functionalities (on the upper part) and the hardware requirements of these functionalities (on the bottom part).

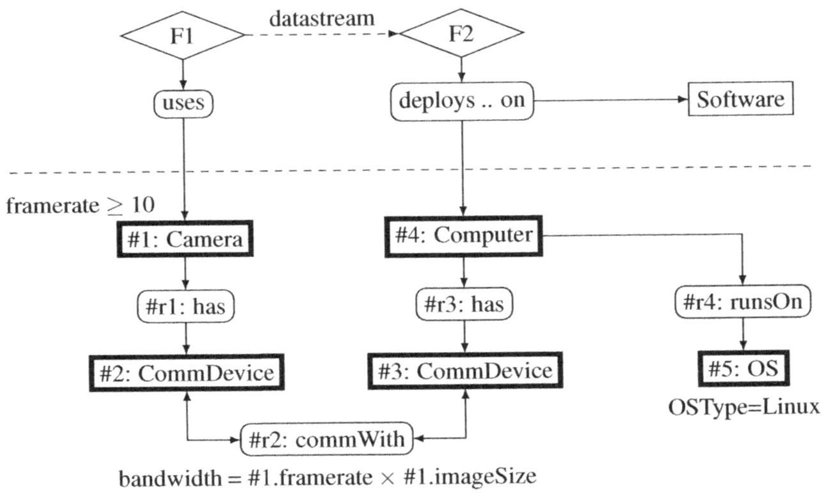

Figure 4.2. *Example of application graph*

4.3.3. *Mathematical formalization of the deployment process*

We model the hardware infrastructure and the applications as graphs. To use an application on the hardware infrastructure, we have to find a graph of the hardware infrastructure that will support the requirements of the application. This projection, respecting the constraints, informs us about which hardware entities to use for the deployment of the application. In this section, we present the mathematical formalism of the deployment process. We first define operators to be able to combine and compare these properties. Then, we introduce the graph homomorphisms as a solution of this projection problem. Lastly, we show that this formalism is extensible and is able to manage various interesting situations.

4.3.3.1. *Properties, nodes and graph*

Each node from the infrastructure graph and the hardware requirement graph of the application graph is characterized by some property values. We define P_i, a set of all the possible values for one property of type i. In example 4.1, the OS property has value Linux, Windows or Mac. The bandwidth property can take any positive value. Framerate of a camera can be a part of $\mathbb{R}^+$, i.e. a set of intervals representing the possible framerates.

EXAMPLE 4.1.–

$$P_{OS} = \{\text{Linux}, \text{Windows}, \text{Mac}, \varnothing, \infty\}$$

$$P_{Bandwidth} = \mathbb{R}^+ \cup \{\varnothing, \infty\}$$

$$P_{Framerate} = \mathscr{P}(\mathbb{R}^+)$$

To each set of property values P_i, we define a binary operation $\vee_i$ and a partial order $\leqslant_i$ to be able to combine and compare properties.

The binary operation $\vee_i$ allows us to combine property values from a same property type. If two nodes of the application graphs are projected on the same node in the hardware infrastructure graph, then we can combine the properties of these two application nodes. This combination informs us about the minimal values required by the hardware node properties to be able to support the deployment. In example 4.2, it is impossible (∞) to use on the same hardware entity two vertices of the application graph which require a Linux and a Windows operating system at the same time. Still, if we want to use two application nodes that require, respectively, 50 and 20 Mbps of bandwidth, we have to find a hardware node with at least 70 Mbps of bandwidth.

EXAMPLE 4.2.–

$$Linux \vee_{OS} Linux = Linux$$

$$Linux \vee_{OS} \varnothing = Linux$$

$$Linux \vee_{OS} Windows = \infty$$

$$50 \vee_{Bandwidth} 20 = 70$$

$$\{5, 10, 15\} \vee_{Framerate} [5, 12] = \{5, 10\}$$

The partial order $\leqslant_i$ on P_i determines if an available property (right) has a sufficient value to match with the desired property (left). In other words, this relation means "is supported by". If the combination of the property values of all vertices in the application graph that are projected onto the same vertex in the hardware infrastructure graph respects the partial order with the property values of this hardware infrastructure vertex, then this vertex can support a part of the deployment. In example 4.3, an application that needs a camera with 5 or 10 images per second can use a camera in the hardware infrastructure that has framerate of 10 images per second.

EXAMPLE 4.3.–

$$Linux \leqslant_{OS} Linux$$

$$70 \leqslant_{Bandwidth} 100$$

$$\{5, 10\} \leqslant_{Framerate} \{10\}$$

The algebraic structure $(P_i, \vee_i)$ forms a commutative monoid with $\varnothing$ as an identity element and ∞ as an absorbing element, ordered by $\leqslant_i$:

$$\forall(x, y, z) \in P_i^3, x \leqslant_i y \Rightarrow x \vee_i z \leqslant_i y \vee_i z$$

Every set of possible values for a property has its own operators that allows us to model both non-consumable properties (operating system type, framerate and software version number) and consumable properties (bandwidth and memory capacity). It is thus possible to make an entity unavailable if it is used by an application already or on the contrary, make it usable by several applications, within the limits of its resources.

Each vertex of the graphs has several property types that characterizes it. The values of these properties can be represented as a vector. We define Π as the Cartesian product of all the P_i sets. Π is the set of the tuples with one value for each property. A vertex is characterized by an element of Π (Example 4.4).

$$\Pi = P_1 \times ... \times P_n = \prod_{i=1}^{n} P_i$$

EXAMPLE 4.4.–

$$\begin{pmatrix} Linux \\ 50 \\ [7, 14] \end{pmatrix} \in \Pi = P_{OS} \times P_{Bandwidth} \times P_{Framerate}$$

We endow Π with a general binary operation $\textcircled{\vee}$ and a general partial order $\textcircled{\leqslant}$ defined by

$$\forall(\pi, \pi') \in \Pi^2, \begin{cases} \pi \textcircled{\vee} \pi' = (p_1 \vee_1 p'_1, \ldots, p_n \vee_n p'_n) \\ \pi \textcircled{\leqslant} \pi' \Leftrightarrow p_1 \leqslant_1 p'_1 \wedge \cdots \wedge p_n \leqslant_n p'_n \end{cases}$$

EXAMPLE 4.5.–

$$\begin{pmatrix} Linux \\ 50 \\ \{5,10\} \end{pmatrix} \circledvee \begin{pmatrix} Windows \\ 20 \\ [7,14] \end{pmatrix} = \begin{pmatrix} \infty \\ 70 \\ \{10\} \end{pmatrix}$$

We define the hardware infrastructure and the applications as $G(V, E, \mathcal{P})$, with V: set of vertices, E: set of pairs of vertices corresponding to the edges of the graph and $\mathcal{P}$: function from V to Π. Each vertex in the graph can be an entity or a relation between two entities which is characterized by a vector of properties in Π. The application $\mathcal{P}$ allows us to get this vector of property values. In the following, "graph" will always refer to such a graph.

4.3.3.2. *Modified graph-homomorphism*

To use applications on the hardware infrastructure, we need to associate with each vertex in the application graph a compatible vertex in the hardware infrastructure graph that meets requirements and respects links. We need to find a morphism from the application graph to the hardware infrastructure graph. We define the function $\phi : G \longrightarrow H$. ϕ as being an enriched graph homomorphism if and only if:

$$\phi : V_G \longrightarrow V_H \begin{cases} \forall (u,v) \in E_G, (\phi(u), \phi(v)) \in E_H \\ \forall y \in V_H, \quad \underset{x \in \phi^{-1}(y)}{\circledvee} \mathcal{P}_G(x) \circledleq \mathcal{P}_H(y) \end{cases}$$

For each vertex in the application graph, ϕ attributes a vertex in the hardware infrastructure graph in which the edges between nodes are respected and the property combination of all vertices in the application graph that are projected to the same vertex in the hardware infrastructure graph respects the partial order with the properties of this image vertex. The set of such morphisms between two graphs represents all the possible projections for the deployment of applications onto the hardware infrastructure.

The set of all homomorphisms between these two graphs $Hom\,(G, H)$ represents all the possible solutions of the projection of an application G on an hardware infrastructure H. The maximum number of possible projections is exponential with the number of application graph vertices:

$$card\,(Hom\,(G, H)) = hom(G, H) \leq |V\,(H)|^{|V(G)|}$$

Finding a solution to the general graph-homomorphism problem is a NP-complete problem [CON 04]. The graph-homomorphism presented in this section is a classical graph-homomorphism enriched with an axiom on the properties.

4.3.3.3. *Formalism extensions*

This formalization allows us to use applications onto a heterogeneous hardware infrastructure, but also to handle different interesting situations. As the applications and the hardware infrastructure are modeled in the same way, we can apply the deployment process from an application to another application. It may be necessary, for huge applications, to make assumptions and reduce the application graph to something less complex. This can be possible as the composition of two morphisms is again a morphism. Let G, H and K be three graphs and ϕ and ψ, two graph-homomorphisms, $\phi : G \longrightarrow H$ and $\psi : H \longrightarrow K$. These morphisms respect the following properties:

$$\forall y \in H, \quad \bigcircledvee_{x \in \phi^{-1}(y)} \mathcal{P}_G(x) \circledleq \mathcal{P}_H(y); \qquad \forall z \in K, \quad \bigcircledvee_{y \in \psi^{-1}(z)} \mathcal{P}_H(y) \circledleq \mathcal{P}_K(z)$$

So, we can write

$$\bigcircledvee_{x \in (\psi \circ \phi)^{-1}(z)} \mathcal{P}_G(x) \circledleq \bigcircledvee_{y \in \psi^{-1}(z)} \mathcal{P}_H(y) \circledleq \mathcal{P}_K(z)$$

If the new less-complex application can be used, then the original one can also be used in the same way.

We may also want to use an application on a virtual machine, and then use this virtual machine on a hardware entity. In addition to the morphism composition property, we can define a function $\phi_* : H \longrightarrow \Pi$, associated with the morphism $\phi : G \longrightarrow H$:

$$\forall y \in H, \phi_*(y) = \bigcircledvee_{x \in \phi^{-1}(y)} \mathcal{P}_G(x)$$

$$\Rightarrow \forall y \in H, \phi_*(y) \circledleq \mathcal{P}_H(y)$$

For each vertex in H, the function ϕ_* associates the minimal vector of properties which can be reduced to allow the deployment of the applications by ϕ. This function ϕ_* allows us to find the minimal configuration of the virtual machine to support the deployment of this application.

We have just seen in this section that we can solve the projection problem by finding a morphism between the application graphs and hardware infrastructure graph. The graph homomorphism definition was extended to include properties attached to each node and operators were defined for each set of values of properties. This defines a new category (in the sense of category theory [MAC 71]).

4.3.4. *Modified graph-matching algorithm*

4.3.4.1. *Existing graph-matching algorithms*

In the previous sections, we have modeled the projection problem for the deployment of applications on a hardware infrastructure for ambient systems as an enriched graph-homomorphism problem. This allows us to reuse one of the existing and well-studied algorithms for graph-matching. In this section, we present different types of exact graph-matching algorithms which are suitable to our problem.

Exact graph matching refers to different variations of the problem. The graph-homomorphism problem is the weaker form of matching, in which only the edges must be respected. It is the one we focus on for our problem. The other graph-matching problems add new assumptions to the general problem. Graph-monomorphism is an injective morphism in which each node in the source graph is projected on a different node in the target graph; graph-epimorphism is a surjective morphim in which each node of the target graph is the image of one or more nodes in the source graph; (sub)graph-isomorphism is a bijective morphism from a graph to a (sub)graph; graph-automorphism is a graph-isomorphism of a graph onto itself and represents a series of permutations of the nodes. The graph K-coloring problem can be considered as a homomorphism from the graph to color to a K-complete graph. Except the graph-isomorphism, all the other problems are NP-complete problems [CON 04]. Consider N_H as the number of vertices in the target graph and N_G, the number of vertices in the source graph. There are potentially $N_H^{N_G}$ different morphism functions to test, and verifying a morphism is in $\Theta(|E_G| + N_G)$.

A lot of algorithms and variants exist to solve these different morphism problems [MCK 14]. Probably the most famous is the one from Ullmann [ULL 76]. It is a branch and bound algorithm which deep-first search the source graph and tries to associate for each node of this graph a node in the target graph, respecting the edges. Backtracking is used when an inconsistent state is reached. Some improvements of this tree-matching algorithm, as the one proposed by Cordela [COR 01], consist of adding an heuristic for the graph exploration. Another algorithm was proposed by Larrosa and Valiente [LAR 04]. In this work, the graph-isomorphism problem was reformulated in a constraint problem which can be solved with a classical CSP engine. An interesting algorithm for graph-morphim problem is the one presented by Messmer and Bunker [MES 00]. It is a variant of a matching technique for expert systems. This algorithm is based on a recursive decomposition of a graph in sub-graphs. In particular, it is efficient to match a graph with a graph database for which the decompositions could be pre-computed. Finally, Babel proposed an algorithm that solves the graph-isomorphism problem in quasi-polynomial time [BAB 15].

4.3.4.2. *Modified branch and bound algorithm for the deployment of applications*

To use an application on the infrastructure, we need an algorithm that finds a homomorphism between the hardware requirement graph of the application graph and the infrastructure graph. We should be able to find all the possible homomorphisms, build them step by step, be able to propose partial solutions and include some heuristics to guide the construction of the solution. Finding an optimal solution is not interesting. We just need to find one of the solutions. The constraints solving algorithms are usually heavy processes that are not suitable for our needs. We promote branch and bounds or gradient descent algorithms. We present in this section, the adaptation of a branch and bound algorithm as the one presented by Chein and Mugnier [CHE 08] that consider the axiom on the node properties.

This algorithm takes as input G: the hardware requirement sub-graph of the application we want to use, and H: the available infrastructure graph. The algorithm deep searches G and tries to successively assign a compatible node from the graph H, following the edges. If there is no solution, the algorithm backtracks until another compatible node can be found in the infrastructure graph. During the execution of the algorithm, the explored nodes from the application graph G and its images in the infrastructure graph H represent a partial graph-homomorphism. A graph-homomorphism is found when all the nodes from the application graph are explored and possess an image. To find the other graph-homomorphisms, we just have to save the current solution and continue the algorithm by backtracking.

Algorithm 4.1 findNextProjection(G,H,projection)

```
input : G, H, projection
output: An homomorphism from G to H
begin
    if projection is undefined then
        projection = initProjection(G,H)
    end
    while step(&projection) and !projection.isFound() do

    end
    return projection
end
```

Algorithm 4.1 is the entry point of matching algorithm. It has been implemented iteratively to be able to runeasily step by step. *Projection* is an object containing two stacks. This object is initialized thanks to the *initPorjection* function. The first stack (*exploration*) allows us to explore the unexplored nodes of the application graph. The second stack (*solution*) associates with each explored node of the application graph, a solution in the infrastructure graph. The *projection.isFound* function tests if all the nodes of the application graph possess a corresponding node on the infrastructure graph.

Algorithm 4.2 step(*projection)

```
input : G, H, projection
output: An homomorphism from G to H
begin
    if !explore(projection) then
        return false
    end
    while !match(projection) do
        if !projection.backtrack() then
            return step(projection);
        end
    end
    return true
end
```

Algorithm 4.2 explores the application graph and try to match a compatible infrastructure node. If there is no compatible infrastructure node, then the algorithm backtracks until it reaches another choice point.

Algorithm 4.3 tries to find an unexplored node in the application graph. If there are no more nodes to explore, the algorithm backtracks on the exploration stack to find another branch of the application graph to explore. If the new selected node already has a solution, the algorithm verifies that the links in the infrastructure graph are correct. Otherwise, the node is pushed in the exploration stack. If both the exploration and solution stacks are empty, the algorithm tries to select a first node to explore thanks to the *projection.selectFirstNode* function. This function can choose a node randomly or thanks to an heuristic. If the exploration stack is empty but not the solution stack, it means that a projection has previously been found. The algorithms invalidate the last match to find the next projection.

Algorithm 4.4 tries to find a compatible node in the infrastructure graph. The *projection.getNextCompatibleNeighbour* returns the next compatible neighbor for the last application node explored, respecting the axiom on the properties of the nodes.

This algorithm allows the user to use one or more applications at the same time. We can consider N application graphs as one non-connected graph with N connected components. Nevertheless, it is possible to use the applications one by one iteratively, by saving the results of the property combinations and by updating the node properties of the hardware infrastructure graph accordingly. With the extrapolation property of section 4.3.3.1, the deployment of new applications increasingly constrains the system. Thus, the removal of one application will reduce some constraints and will not question the running of the others. They will retain their quality of service. However, a solution with a better quality of service may appear. The complexity of the removal of one application is linear with the number of nodes of the application graph in the worst case. If the objective is to adopt optimal solutions, we again have to compute the solution of the deployment of all the remaining applications.

Algorithm 4.3 explore(*projection)

```
input  : A pointer on projection object
output: true is there is another application node to match
begin
  if projection.exploration.empty() then
    if projection.solution.empty() then
      sourceNode = projection.selectFirstNode(); if undefined == sourceNode
      then
        | return false
      end
    else
      | projection.invalidateLast() return true
    end
  else
    repeat
      sourceNode = projection.getNextUnexploredNode() if undefined ==
      sourceNode then
        | projection.exploration.pop()
      else if projection.hasImage(sourceNode) then
        if projection.H.hasLink(projection.getImage(projection.exploration.last()),
        projection.getImage(sourceNode)) then
          | sourceNode = undefined
        else
          | projection.invalidateLast()
        end

    until undefined == sourceNode;
  end
  projection.explore(sourceNode) return true
end
```

Coming back to our example (Figures 4.1 and 4.2), the algorithm chooses a node in the application graph and tries to find a compatible node in the infrastructure graph. Here, we choose Camera 1 of the application graph as the starting point. Cameras 1 and 3 of the hardware infrastructure are compatible with this node because they respect the partial order between their properties. We choose to assign the hardware entity Camera 1 to continue the algorithm. In Figure 4.3, we show the exploration tree of this solution to the deployment algorithm. We see that computer 5 in the hardware infrastructure cannot be used because of the operating system property (Windows) which does not respect the partial order with the property of the corresponding application node (Linux). So, the algorithm backtracks and invalidates its assignations until it goes back to its last choice point: the comWith #r2 of the application graph. The algorithm chooses to assign a new entity and explores a new branch. The process is repeated until a solution is found. Note that the exploration tree in Figure 4.3 gets all the possible solutions and does not stop to the first one. It is also interesting to note that, in this example, the algorithm automatically configured the frame rate of the cameras for each solution. In the first solution, the frame rate of Camera 1 is limited to 12 or 24 images per second because of the bandwidth property between the camera and the computer hosting the processing software. Similarly, the frame rate of the camera for the second solution should be 12 images per second for the same reasons.

Algorithm 4.4 match(*projection)

```
input  : A pointer on projection object
output: true is a compatible node is found
begin
    targetNode = projection.getNextCompatibleNeighbour() if nodeH == undefined
    then
        projection.solution.last().match(null)
    else
        projection.solution.last().match(targetNode)
    end
    return (nodeH != undefined);
end
```

4.3.5. *Conclusion*

AmI systems are characterized by a strong heterogeneity of hardware and software. This heterogeneity increases the complexity of the automatic deployment of new applications. To facilitate this kind of deployment, we consider that the entities of the hardware infrastructure and the software entities of the applications must be decorrelated. This allows the user to be flexible for the addition or removal of hardware entities, or for the deployment of new software applications onto an existing infrastructure.

In this section, we proposed a graph-based model to describe, on one side the hardware infrastructure, while the software applications we want to execute on the other. From this model, we presented an algorithm that projects a set of applications onto the hardware infrastructure. It determines a set of compatible hardware entities to run each functionality of the application. The algorithm also allows the setting of the hardware entity characteristics to satisfy expressed needs in applications. Supported hardware properties include non-consumable ones, as OS type, as well as consumable ones, as network bandwidth. When an application requirement is projected onto an element of the hardware infrastructure, some of its properties can automatically be configured to fulfill the requirements. The selection or the removal of an application can be done dynamically because the algorithm updates the properties of the hardware infrastructure in terms of resources used by applications. The mathematical formalism describes a category (in sense of category theory) where the objects are systems, modeled by graphs whose nature (hardware, software or virtual) does not matter and where the morphisms model the deployment.

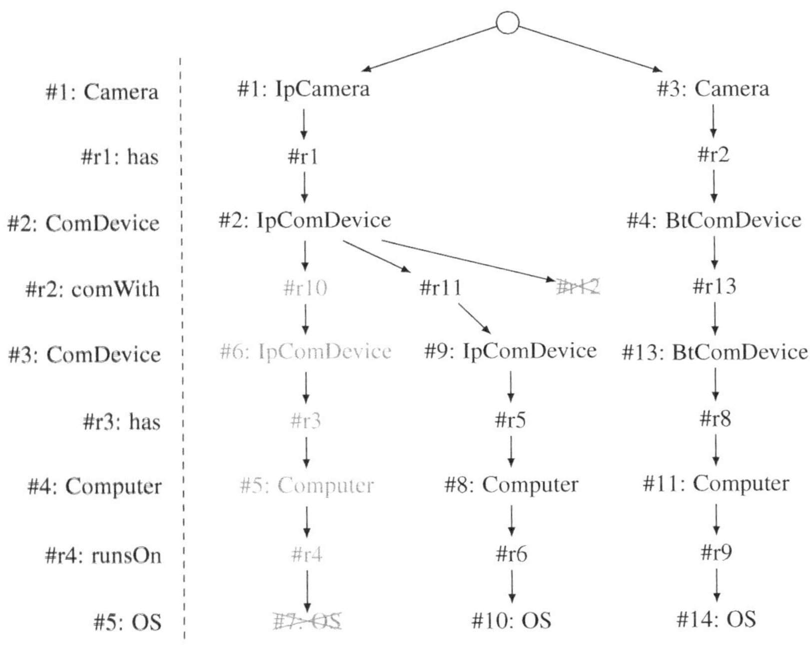

Figure 4.3. *Exploration tree of the deployment algorithm*

4.4. Multi-agent middleware for ambient systems

In the previous section, we presented a way to model the hardware infrastructure and the applications for ambient systems using graphs. We proposed to reason on these graphs to find in the infrastructure the entities that support the running of these applications by using a graph matching algorithm. This allows us to take advantage of the hardware heterogeneity to use ambient applications. However, to manage cities, we need much more than a knowledge representation. We have to include these reasoning and deployment mechanisms in a middleware that is also able to handle other points like privacy, scalability, robustness and others.

We present in this section a scenario that illustrates the deployment of application. From this scenario, we explain why multi-agent systems are well adapted to design the deployment software and ensure privacy. We detail the multi-agent structure and at last provide details on the implementation, using a goal-oriented approach.

4.4.1. *Scenario*

The scenario we used in this chapter highlights both the dynamic deployment of distributed applications and privacy management encapsulated in both the agents and agent organizations.

Mr Snow uses a *video doorkeeper* for dependant persons (e.g. visually impaired) application at his home. When someone rings at the door, the image of the entrance camera is displayed on a screen near Mr Snow, making sure he can properly see the person. He can, then, talk with the person and decide whether or not to remotely open the door.

It is a Saturday morning and Mr Snow is waiting for a parcel that is likely to be delivered to his home by the courier at any moment. While he is grooming himself in the bathroom, his neighbor, Mr Den, rings the door. The smart house, aware that Mr Snow is in his bathroom, selects the connected mirror of the bathroom, instead of any of the other display screens of the house, as a support to display the image stream of the entrance camera. Mr Snow, not being able to receive his guest, informs him, thanks to the microphone in the mirror, that he will meet him in an hour. Before going to see Mr Den, Mr Snow has to go shopping two blocks from his house. During his trip, he receives a notification on his smartphone: somebody rings at his door. Based on the lack of location information, the deployment software decides to use his smartphone as a default terminal for the doorkeeper application. The doorkeeper application is automatically used to display the image stream of the camera directly to Mr Snow's smartphone. It is his brother who came to bring him back some CDs he had borrowed last week. After having informed him that he was not at home, he remotely unlocks his entrance door so his brother can leave the CDs on the table. After shopping, Mr Snow goes to his neighbor, as he does every Saturday. In the middle of their conversation, he is notified that someone is ringing his door again. This time, it is an unknown man. He decides to display the image of this man on Mr Den's television to ask him if he recognizes the guest. By default, Mr Snow does not have the right to use any devices that he does not own, but as he and Mr Den are good friends, the latter has authorized him to access the television when he is at home. The doorkeeper application is reused dynamically to use the requested hardware entities. Neither Mr Snow nor his neighbor knows the visitor. Mr Snow decides to activate the microphone of the camera to talk with the man. After a short talk, he learns that the unknown person is the expected courier. He tells him he will be right back and asks him to wait a minute. Mr Snow thanks Mr Den for the coffee and goes back to his house to receive his parcel in person.

The important point in this scenario is not the video doorkeeper application, but the way it is used dynamically in the environment, considering the user's context. The scenario shows three deployment situations. First, the application was used for the use in the user's own home infrastructure, but in a less usual place: the bathroom. Then, the application was used on the phone as a default solution, because the location of the user could not be found. Finally, the application was used on the infrastructure of another user, given that the necessary access rights had been granted.

To achieve the deployment of any applications on an existing hardware infrastructure and to handle privacy constraints, we use the MAS features. The next section presents why MAS paradigm is suitable for our purpose.

4.4.2. *Multi-agent modeling*

Our scenario highlights several necessary specificities of the deployment software. This software has to dynamically deploy and undeploy distributed AmI applications in an environment that is also dynamic: when a visitor rings the doorbell, the deployment of the video doorkeeper should start, considering the available hardware entities and the location of the user, to choose the most relevant screen for displaying the image of the camera. Given its distribution and openness that characterize the AmI domain, privacy is a very important characteristic of the deployment software. In this scenario, we focus in particular on *resource privacy*. Mr Snow is the owner of the hardware entities in his house and he does not want unauthorized people using or even knowing of the existence of these resources. Finally, autonomy and robustness of the system are also very important specificities: if my neighbor's system failed, mine should continue to work normally and should not be affected.

As the required software demands distribution, privacy, context management, autonomy and robustness, we identified MAS as a suitable solution. Through its modularity, this paradigm facilitates a local processing of the data and guarantees the autonomy of the different parts of the hardware infrastructure, thus handling aspects of privacy and robustness. To solve the dynamic deployment problem, we use the graph representation for the hardware infrastructure and the requirements of the usable applications, and the graph matching algorithm, presented in the previous section, to find the entities that can support the running of the application.

In the next sections, we present the modeling of agents and the agent organization for our deployment solution, while focusing on the encapsulation of resource privacy.

4.4.2.1. *Agents and artifacts*

In this section, we present the two kinds of components of the deployment software. The system is made up of different agents and artifacts interacting with each other. The agents are able to reason on the deployment of applications with respect to the available infrastructure, whereas artifacts are resources and tools that can be instantiated and/or used by agents in order to interact with the environment [RIC 03, RIC 11]. The former are autonomous and goal-directed whereas the latter are not.

We propose four classes of agents: *Infrastructure Agents*, *Infrastructure Super Agents*, *Application Agents* and *User Agents*. These classes allow us to have a clear separation between the infrastructure, the applications and the users interacting with the system. For each class of agents, we identify the main goals:

– an *Infrastructure Agent* deals with a part of the global hardware infrastructure. It uses the graphical representation of the infrastructure available (hardware entities, relations and properties). This graph representation is never shared with other agents. The *Infrastructure Agent* reasons on it to propose solutions for the deployment of applications, thanks to a graph-matching algorithm. This class of agents has several functions, as it has to 1) keep the infrastructure graph up to date; 2) propose solutions for the deployment of applications, considering the available hardware infrastructure, but also the sharing and privacy policy; 3) deploy or 4) undeploy functionalities of an application;

– an *Infrastructure Super Agent* is a representative of a set of *Infrastructure Agents* which are related to if, forming a *group*. This class of agents acts as a proxy between the agents inside and outside of the group;

– an *Application Agent* manages an entire application during its runtime. It has a graph-based description of the application. This application graph describes the functionalities of the application and the hardware requirements to be fulfilled to use these functionalities. The main goals of this class of agents are to 1) guarantee the consistency of the application 2) deploy or 3) undeploy functionalities of the application if necessary. The *Application Agent* has to interact with several *Infrastructure Agents* to use the functionalities of the application over the infrastructure.

– the *User Agent* is the interface between the user and the other agents of the deployment software. Through this agent, a user can 1) request the deployment or 2) undeployment of applications.

In addition to these four classes of agents, we also propose two classes of artifacts:

– first there are *deployment artifacts* [FLI 08] that can be used by the *Infrastructure Agents* to use some parts of an application on the hardware entities or configure these hardware entities effectively so that they can be used by the application.

– the second class of artifacts are the *functionalities* of the applications themselves. Some of them can also provide useful contextual information to the deployment of software (location of the users, available bandwidth etc.) to help the agents to keep their application or infrastructure graph up-to-date.

In the video doorkeeper scenario, there are three *Infrastructure Agents*. The first one manages the hardware entities located in the living room of Mr Snow, like the television. The second one manages the entities of the bathroom like the connected mirror and the last one manages the house of the neighbor. We also find two *Application Agents*. The first one manages the video doorkeeper application; when a visitor rings the doorbell, this *Application Agent* triggers the deployment of the video interaction functionality. The second one manages the application that provides the location of the Mr Snow inside his own house to his own *Infrastructure Agents*. The contextual location information is useful for using other applications. Indeed, the display screen of the video doorkeeper application has to be chosen near the user. Then, we have two *User Agents*. The first one is the interface between the deployment software and Mr Snow and the second one is owned by Mr Snow's neighbor. At last, we have a certain number of deployment artifacts that can configure the display screens, the cameras or use software on devices (TV box, connected mirror, etc.).

The agent decomposition encapsulates a part of the privacy mechanism. The graphical representation of the available hardware infrastructure managed by an *Infrastructure Agent* is only known by this agent and is never shared with others. Moreover, the architecture used helps keeping a clear separation between the applicative part, managed by the *Application Agents* and the hardware part, monitored by the *Infrastructure Agents*. As agents only have a local view of the system, the privacy is enhanced. In the next section, we show how using this agent organization improves resource privacy and allows the definition of privacy policies for the use of these resources.

4.4.2.2. *Multi-agent organization*

All the agents presented in the previous sections cannot interact freely. They are constrained by the way we decide to organize them and the privacy policy we define. *Infrastructure Agents* can be grouped behind an *Infrastructure Super Agent* which, as stated before, acts as a proxy for the agents of the group. From an outside view, this *Infrastructure Super Agent* is seen as a normal *Infrastructure Agent*. In our scenario, the living room and the bathroom *Infrastructure Agents* of Mr Snow are grouped behind an *Infrastructure Super Agent* representing the house of Mr Snow. Similarly, the *Infrastructure Agent* managing the house of Mr Snow's neighbor is a super agent, regrouping several *Infrastructure Agents* (or other sub super agents). The advantage of such organization is that it is easy to abstract groups of agents and make them invisible from the outside, resulting in a multi-scale property that helps us improve privacy. Mr Snow knows about his own *Infrastructure Agents* (bathroom and living room), but he does not have to know anything about the details of Mr Den's *Infrastructure Agent* organization. If he wants to interact with his neighbor's house, he has to know Mr Den's *Infrastructure Super Agent* and have the required access rights granted (as described below). The upper part of Figure 4.4 shows the organization of the *Infrastructure Agent* from Mr Snow's point of view. This kind of organization can be implemented with both a hierarchy or a holarchy. In a hierarchical organization [FOX 81], *Super Agents* are represented by a software agent that acts as a proxy between the agents of the group and the outside. This agent is the favored interface between the group and the outside. In a holarchy [KOE 67], a super agent is not a concrete software agent but is represented by the sum of all *Infrastructure Agents* of the group. Each of these agents can be a representative of the super agent, unlike in a hierarchy where there is only one representative that may become a single point of failure. A holarchy is more complex to implement, but is more flexible too and can evolve dynamically. Some sophisticated hierarchies are similar to holonic organizations [HOR 04]. In our implementation, we used hierarchies because they are easier to implement and debug, but both models could have been implemented and coexisted in our system.

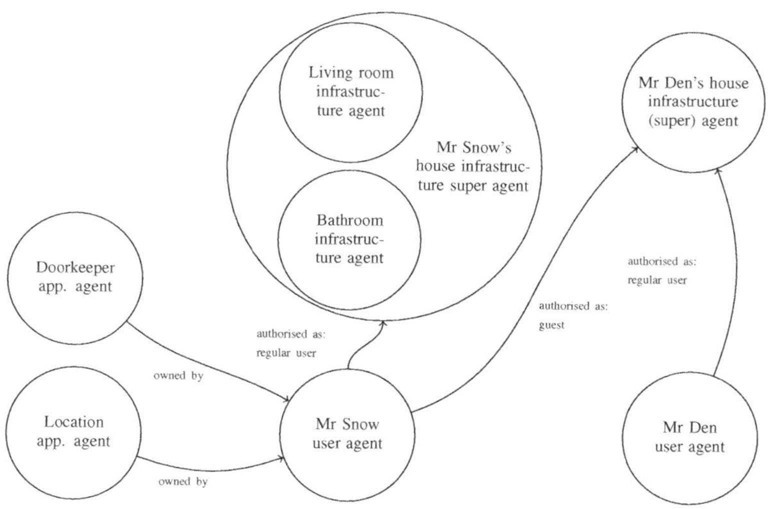

Figure 4.4. *Example of a basic application graph*

The organization of *Infrastructure Agents* ensures privacy by hiding information about the structure of its sub-organizations. However, to improve privacy by controlling the use of resources, we also propose to use sharing policies. *User Agents* can be authorized, by the owner of some hardware infrastructure, to use some parts of its infrastructure, and cooperate with the associated *Infrastructure Agents* or *Super Agents*, to use applications. If a *User Agent* is not known and is authorized by the *Infrastructure (Super) Agent*, it cannot use the hardware resources proposed by this agent. If a *User Agent* is known, it can have different authorization levels. For example, (1) Admin level: the agent (and implicitly its user) has full access to the resources proposed by the *Infrastructure Agent* and can reconfigure the *Super Agent*, manage the authorization levels, add or remove *Infrastructure Agents* to the *Super Agent*; (2) Regular user: the agent has access to the resources of the *Infrastructure (Super) Agent* but it cannot reconfigure authorization levels or agent organization; (3) Guest: the agent has a restricted access to the resources. Only the resources considered as non-critical by an administrator are allowed to be shared. These authorization levels can be modified by the administrator of the *Super Agent*. In the video doorkeeper scenario, Mr Snow's *User Agent* is a Regular user for his home *Infrastructure Super Agent*, but it is just a Guest to his neighbor's home *Infrastructure Super Agent*. As such, it has only access to the television of Mr Snow's neighbor. This allows us to ensure privacy of the other resources of the neighbor. At last, the *Application Agents* are created by *User Agents* and depend on them. These *Application Agents* have the same authorization level as the *User Agent* creator. They can interact with the authorized *Infrastructure Agents* to use their applications effectively. Figure 4.4 shows the agent structure of the doorkeeper scenario; the *Infrastructure Agent* organization, the authorization level of the *User Agents* and the *Application Agents* that are bound to their *User Agent* creator.

In this section, we have shown how privacy is preserved through encapsulation in our MAS. *Infrastructure Agents* keep the information about the hardware infrastructure secret. The *Infrastructure Agent* hierarchy keeps the details of the agent organization hidden. Privacy policies can allow or prevent the sharing of resources to *User Agents*. This results in privacy by design.

4.4.3. *Distributed reasoning*

The algorithm presented in section 4.3.4.2 allows the user to find the projection of applications on an infrastructure. However, it is centralized and does not take into account important specificities of smart environments, like privacy or scalability. We have seen in the previous sections that our MAS middleware tackles this problem. However, each *Infrastructure Agent* has to deal with a part of the global hardware infrastructure that remains private to ensure privacy. Thus, we have to distribute the representation of the infrastructure graph and adapt the previous algorithm into a decentralized version to solve the projection problem.

4.4.3.1. *Graph representation modifications*

The main purpose of the MAS is to deal with resource privacy by distributing the hardware infrastructure representation to several agents. Each infrastructure agent handles a graph of the global infrastructure graph. The point is to know where it is relevant to cutting the graph and isolating the sub-graphs. Some parts of the graph cannot be separated. This is the case for an entity as a computer and its dependencies (RAM, CPU, communication devices etc.). These form a non-dissociatable graph. These graphs are linked through their mutual nodes. Two graphs linked together are adjacent. The way we decide to cut the global infrastructure graph into graphs depends on the application field. For our smart home demonstrator, we decided to cut the graphs at the level of the communication links between two hardware entities and more specifically at the network nodes. So, the graphs are linked together by these different networks that are the ones to be shared. Figure 4.5 presents an example of a global infrastructure graph. The filled nodes represent the nodes where the graph can be cut (e.g. network nodes). The decomposition of this graph into sub-graphs is illustrated in Figure 4.6. Each graph, thus, obtained can be handled by an *Infrastructure Agent* that never shares this representation and thus keeps the hardware infrastructure secret.

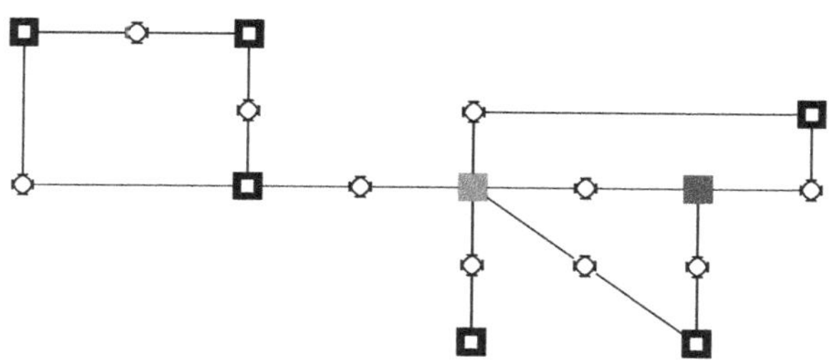

Figure 4.5. *A global infrastructure graph*

With this decomposition of the global infrastructure graph, we can apply the previous algorithm to one graph. This one tries to find a solution to the projection of the application locally. If there is no local solution, then a partial solution is generated and the algorithm is reexecuted on an adjacent graph to complete the solution. The global algorithm stops when a complete solution is generated or when all the graphs have been explored without modifying the partial solution.

Back to our scenario, the Mr Snow has two infrastructure agents managing his hardware. The first one handles hardware entities from the entrance hall while the second, from the living room. In the entrance hall, there is a door that can be opened remotely connected to a ZWave network; a doorbell using the EnOcean technology, informing if a guest rings the doorbell and an IP camera connected to the local IP network of the house. In the living room, there is an open source TV box connected to the family television integrating an IP interface and an EnOcean controller and a multi-touch coffee table managed by a microcomputer connected to both local IP and ZWave networks. The hardware infrastructure is represented by the following two graphs (Figures 4.7 and 4.8). For readability, the node properties are not represented on these graphs. The filled square represents the IP Network; the filled circle, the ZWave network; and the non-filled circle, the EnOceanNetwork. Each graph is known by only one agent.

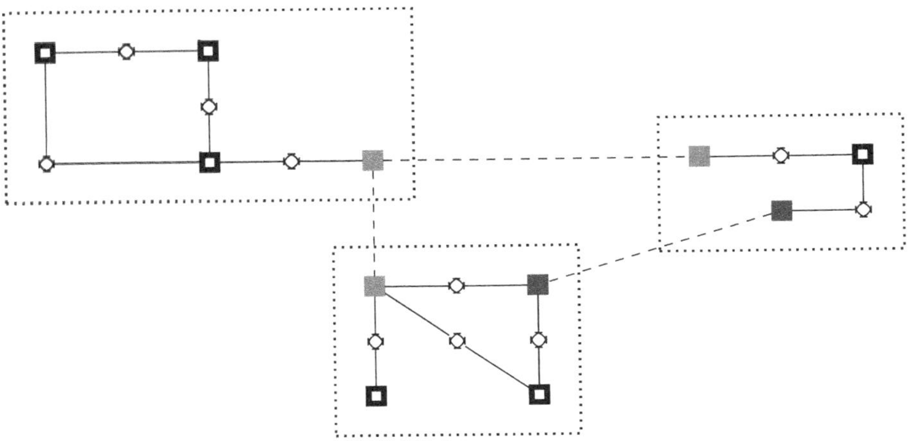

Figure 4.6. *Sub-graph decomposition*

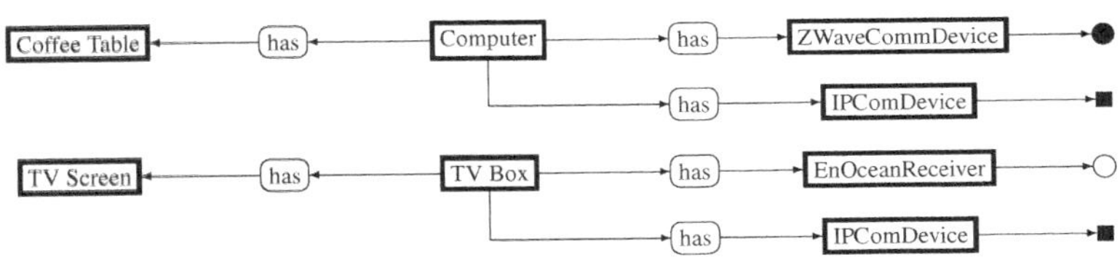

Figure 4.7. *Living room infrastructure graph*

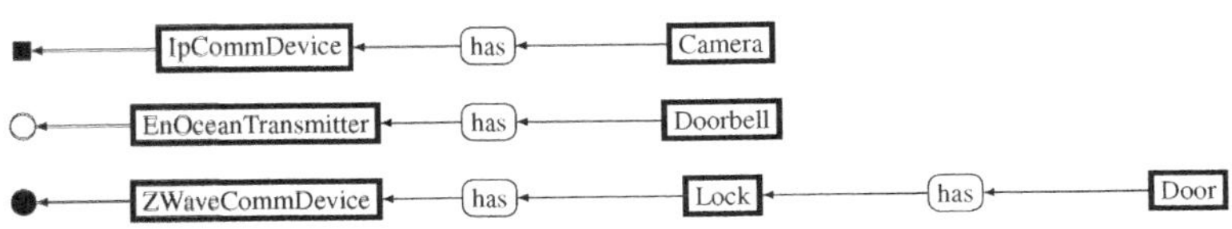

Figure 4.8. *Entrance hall infrastructure graph*

4.4.3.2. *Agent cooperation*

There are, of course, several ways to interact with the other *Infrastructure Agents*, depending on whether we would like to search depth-first, like the centralized algorithm or breadth-first. Algorithm 4.5 shows the breadth-first version. This version returns all the projections.

Algorithm 4.5 breadthFirst(G, H, projection)

```
input : G, H, projection
output: A set of homomorphisms from G to H
begin
    allSolutions = [] newProjection = findNextProjection(G, H, projection) if
    newProjection.isFound() then
        allSolutions.push(newProjection)
    else if projection!=newProjection then
        foreach Hi adjacent to H and H != Hi do
            allSolutions.concat(breadthFirst(G, Hi, newProjection))
        end
    return allSolutions
end
```

It is also possible to do this asynchronously and stop the algorithm when one solution is found. To adapt this algorithm to our agents, we just need to ask each adjacent *Infrastructure Agent* to solve the projection problem by using the graph-matching algorithm on his own infrastructure graph. The partial solution, shared among the agents, indicates only the associated node without revealing the structure of the infrastructure, thus ensuring the privacy.

The deployment of an application is requested by the user. It is always the user agent of the user that gets the application graph and starts the cooperation. As the personal agent does not manage any hardware entities, it has to cooperate with deployment agents. The first problem is to determine the agents with which it will cooperate first. This can be done thanks to different methods (confidence indicator, overload indicator, clustering and authorization levels, various heuristics etc.). For some applications, several hardware nodes have to be associated by the user itself. That is the case for critical entities that cannot be selected automatically or require the user approval for more security (like a doorlock or a camera). The personal agent will ask the deployment agents or clusters to get information about some hardware entities. If it has the right credentials, the deployment agent informs about its available entities. When some nodes are already associated, the selection of the first agent with which to cooperate is more easy.

4.4.4. *Design and implementation*

The agents were designed using a goal-based model due to its benefits to the autonomy and robustness of the application [CHE 05]. Goals are specified by describing their associated plans: higher level *goal plans* describing relationships between goals and lower level *action plans* for concrete actions. This goal-based representation is based on the goal-plan separation (GPS) approach [CAV 14], where each agent has a main goal plan (i.e. plan without any actions, so only decisions, perceptions and goal adoptions) that describes the top level behavior, which can be pursued using other goal plans or directly action plans (i.e. plan without any goal adoptions). This approach helps us handle agent complexity through a multi-level description, from top level abstract behaviors with goals to concrete action plans. Using goal-plans also has the advantage of specifying the relationships between goals in a plan format.

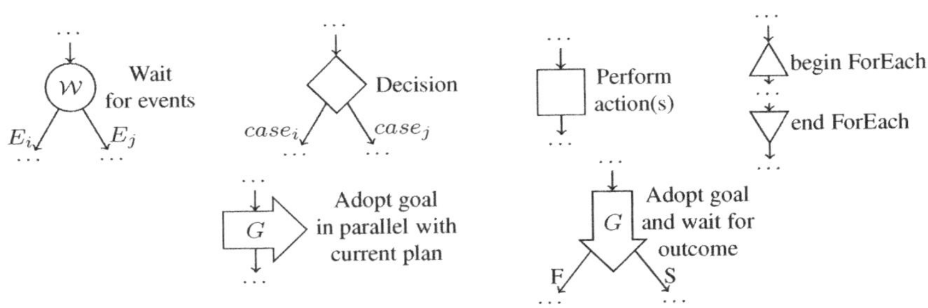

Figure 4.9. *Flowchart nodes for efficiently describing the plans of goal-driven agents*

Plans are represented using a flowchart notation we adapted for modeling goal-driven agents (Figure 4.9). The notation contains the main elements that allow for the behaviors of agents to be defined: decisions. For this application, we considered a simple goal model (similar to a *perform* goal [BRA 05b]) where a goal is successful ("S") when the plan executing for it ends with "End ok". This allowed us to keep a simple goal lifecycle appropriate for using in our application, while still gaining from the features of the goal-based design.

We continue by describing in detail the agents of the system. Since the *Infrastructure Super Agent* is only a proxy between the agents of the group it represents and the other agents outside this group, its implementation is not detailed here [AU 1]. In what follows, P_{Xi-j} are the plans for a goal G_{Xi}.

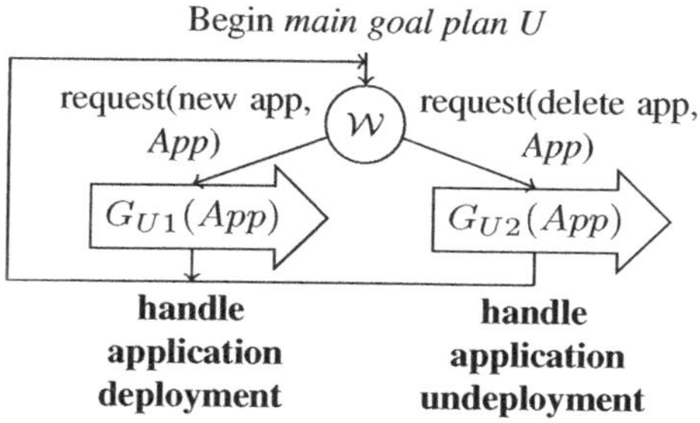

Figure 4.10. *User Agent: main goal plan*

4.4.4.1. *User Agent*

The *User Agent* acts as an interface between the user and the deployment MAS. The goal plan of the *User Agent* (Figure 4.10) waits for user input and, depending on the received request, adopts the necessary goal, corresponding to the agent functions identified in section 1.5.2.1. The plans of G_{U2} and G_{U1} are similar: they create an *Application Agent* or request an application not to be used, wait for a confirmation and display the information to the user. The *User Agent* also allows changing the privacy policies and settings, but this was not represented here.

4.4.4.2. *Application Agent*

The *Application Agent* is created by a *User Agent*. It tries to use a precise application by cooperating with one or more known *Infrastructure (Super) Agents*, from which it does not need to have any infrastructure details.

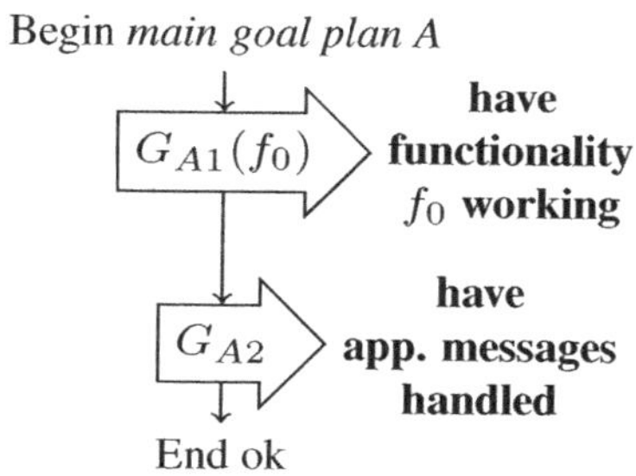

Figure 4.11. *Application Agent: main goal plan*

Once created, an *Application Agent* adopts two goals (Figure 4.11): G_{A1} for using an initial functionality (Figure 4.12) and G_{A2} that waits for internal events for new deployments or undeployments (Figure 4.13). The deployment is done in two steps: first the agent obtains a deployment solution from *Infrastructure Agents* via G_{A3}, and then it requests the deployment according to this solution through G_{A4}. The *Application Agent* sends a list of the requirements described in the application graph to the *Infrastructure Agent* and the solution it receives contains the list of requirements that could be fulfilled. Note that the reply does not contain any actual infrastructure details, which is important for the privacy of the infrastructure. It can be seen (Figure 4.14) that the agent may need to call multiple *Infrastructure Agents* to obtain a complete deployment solution. An *Infrastructure Agent* tries to find in its own infrastructure the hardware entities that match the requirements of the application. However, if these requirements only partially match, the *Infrastructure Agent* will return a partial solution to the *Application Agent*. In this case, the latter will call another *Infrastructure Agent* that will continue to match the requirements of the application. Once a solution has been found, the *Application Agent* interacts again with the concerned *Infrastructure Agents* to effectively use the functionalities of the application: plan P_{A4-1} simply sends a message and waits for a confirmation.

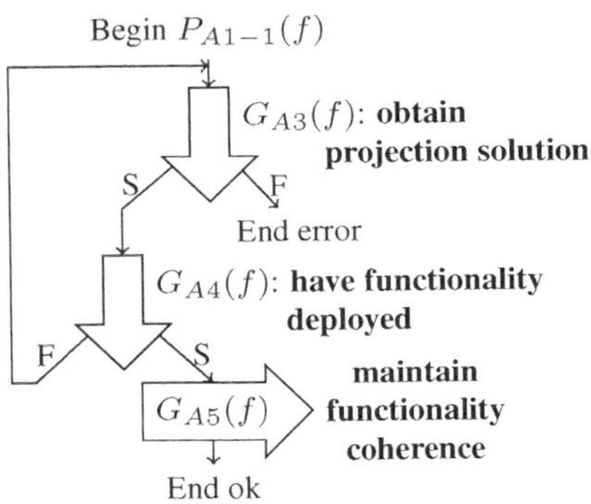

Figure 4.12. *Application Agent: goal plan for G_{A1}: "have functionality f working"*

After a functionality was used, the agent monitors it through G_{A5} (with its plan in Figure 4.15) to adapt the deployment to the current context: infrastructure inconsistency (e.g. changing infrastructure availability and changing user location) and messages from the application itself (e.g. new guest at the door). An application message can result in multiple requests for deployments and undeployments. Internal events are used to control the execution of different plans. Deploy and undeploy events originate in the plan for G_{A5} and trigger the adoption of G_{A1} for the deployment of other functionalities or redeployment of the current one and G_{A6} for the undeployment of the functionality. As each functionality is monitored by an instance of G_{A5}, in the case of an undeployment, the plan of G_{A6} signals the corresponding G_{A5} to stop through a *kill* event (besides sending a request message to the corresponding *Infrastructure Agent*).

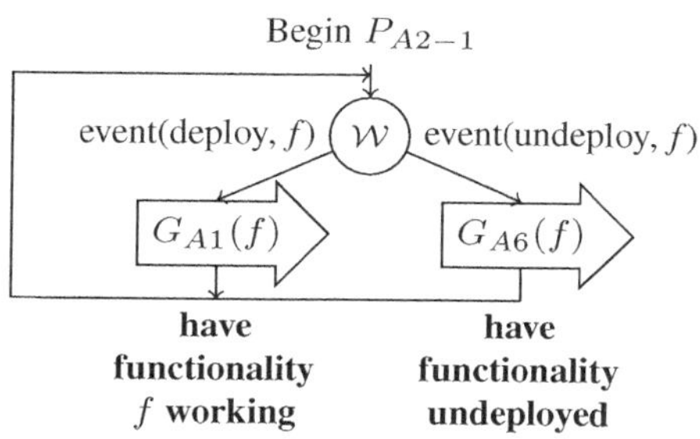

Figure 4.13. *Application Agent: goal plan for G_{A2}: "have app. messages handled"*

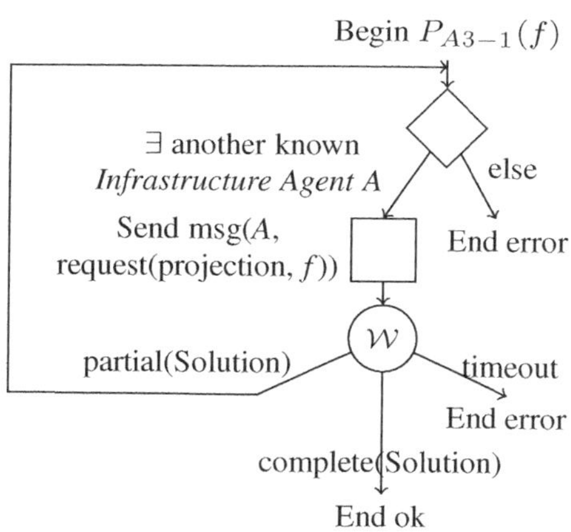

Figure 4.14. *Application Agent: plan for G_{A3}: "obtain projection solution"*

Note here that the *Application Agents* only handle the application deployment. The application itself is in charge of its own actions, data and privacy.

4.4.4.3. *Infrastructure Agent*

An *Infrastructure Agent* receives requests from *Application Agents* that it tries to satisfy (Figure 4.16). Only requests originating from known *User Agents* are treated i.e. only applications from agents that were granted one of the levels of authorization are accepted.

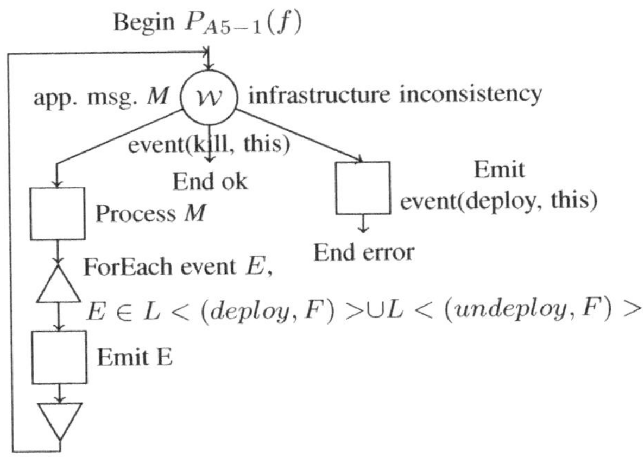

Figure 4.15. *Application Agent: plan for G_{A5}: "maintain functionality coherence"*

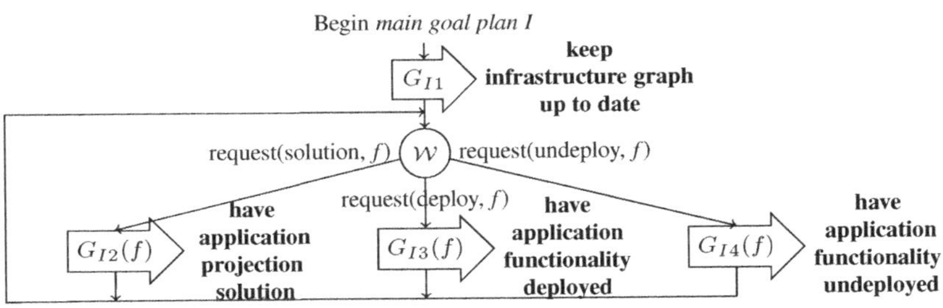

Figure 4.16. *Infrastructure Agent: main goal plan*

When it receives a request for a deployment solution, the *Infrastructure Agent* uses the graph matching algorithm to determine if it can fulfill the requirements of the request (Figure 4.17) using the devices it manages. The algorithm takes into consideration the levels of authorization of the involved *User Agents*. If it cannot produce a complete solution, the *Infrastructure Agent* requests the help of other agents in its group, but without informing the *Application Agents*. In this way, the components of the infrastructure remain private. If a complete solution is eventually produced and the *Infrastructure Agent* is given the order to use the application, it will dispatch the deployment tasks to its own deployment artifacts as well as to any other *Infrastructure Agents* that were included in the final solution. In case any of these requests fails (e.g. an artifact malfunctions), the whole application is undeployed and the *Application Agent* is informed, which will cause it to restart the deployment procedure.

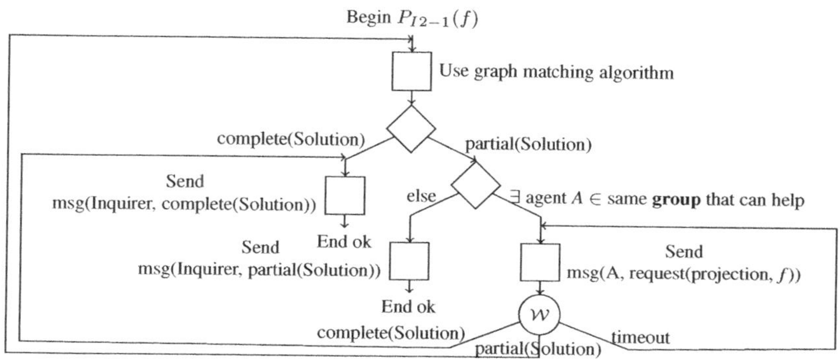

Figure 4.17. *Infrastructure Agent: plan for G_{I2}: "have application projection solution". "Inquirer" can be an Application Agent or another Infrastructure Agent*

In parallel with the request handling, the agent also adopts G_{I1} which listens for agent and artifact information to manage the graph the devices corresponding to the *Infrastructure Agent*. In the case of an inconsistency (e.g. Mr Snow leaves Mr Den's home, so any display he used there are no longer relevant for the application), the agent informs the *Application Agents* that it will need to reuse the concerned parts of their applications.

4.4.4.4. *Implementation*

A demonstration model of the deployment software has been developed in an apartment replica attached to our laboratory. This home replica implements various scenarios applied to home care for dependent persons, including the presented scenario. These scenarios are using commercial connected devices tweaked to be horizontally connected, thanks to the deployment software. These demonstrations run continuously and can be tested by any visitor.

Our goal is to run the MAS on different devices like smartphones or embedded systems with few resources. Most of existing regular MAS platforms like Jade for instance are memory-consuming and Java-oriented platforms [KRA 15]. They are not suitable for our purpose. That is why we designed our own MAS platform in JavaScript. Indeed, Web technologies are fully interoperable and the agents can easily be run on devices like smartphones or the Raspberry Pi. Visualization and interfaces are also JavaScript Web applications. The agents embed a monitoring and debugging Web server that proposes interfaces for interacting with it. The effective deployment is handled by deployment artifacts. The demonstration model handles *ssh* and *puppet* artifacts to use and run software on UNIX systems (computers, micro-computers, Unix-based devices, etc.). We also implemented a specific deployment artifact that configures the frame rate of IP cameras. In this implementation, we mostly used IP devices. We also integrated EnOcean devices. These devices, however, are handled by a hard-coded gateway that extends the IP network to EnOcean devices. The next stage will be to handle multiple means of communication by automatically using gateways or proxies between the devices when needed. At last, the agent implementation was, in first place, not obvious. The multi-level GPS approach made it intuitive to develop.

This realization helps us to figure out the difficulties of handling the heterogeneity of hardware entities. We are now able to handle applications through an AppStore for Smart Homes. These applications can be automatically used in a real environment, using the available hardware devices, and including mechanisms to ensure privacy management of the resources. This provides a concrete base for the implementation of a complete middleware for the deployment of distributed applications in a smart environment.

4.5. Conclusion

The technological improvements and the ongoing miniaturization of electronic devices allowed the establishment, on one side of the hardware, of network and energetic infrastructures able to support ubiquitous applications, while on the other side, the context-aware and intelligent applications that are designed to help users in an unobtrusive way. The Internet of Things and ambient intelligence are domains that will be the catalyst of the next revolution in information technologies. The former focuses on the infrastructure level and the second one on the application level. However, we have shown there is still a gap between the hardware infrastructure and the intelligent applications: nothing exists to allow the reasoning on the heterogeneity of these ambient systems nor on the deployment of applications on the existing infrastructure.

We proposed a model for ambient systems that are capable of solving this problem: a graph representation for infrastructure and the applications, and a mathematical formalization for the deployment process of these applications on the hardware infrastructure. The proposed algorithm finds, for each requirement of the application we want to use, the compatible hardware entities that are available in the infrastructure. These entities found are configured automatically to be the support for the execution of the application.

The use of MAS has made it possible to introduce privacy measures at the architecture and organization level, on top of which we were able to add a user-defined privacy policy mechanism. This was an important criterion for the choice of the agent paradigm since in the domain of ambient intelligence, there are often different infrastructure owners that need to ensure the privacy of their resources. The separation between the applicative and the infrastructure layers, together with the decentralized approach also enhance the robustness of the solution. The clearly delimited entities, with either virtual (the applications) or physical (users, infrastructure elements) correspondents, guided the agentification. The use of a goal-based representation for agents together with the goal-plan separation approach facilitated the modeling task. The specific plan notation was efficient in describing the agent plans both during design and for presentation purposes.

This chapter aims to ease the use and the communication between connected objects. Promoting this horizontal communication, we favor direct exchanges and a local processing of data. This allows us to find mechanisms that guarantee the resources and data privacy. This point is not easy with a vertical system, like the majority of current industrial systems.

In terms of future work, for the deployment software, data privacy in the used applications should also be taken into consideration in addition to the resource privacy discussed here. We would like to facilitate the local processing and storage of the data by defining data privacy policies which should be facilitated by the modularity of the MAS. The user should decide which kind of data he authorizes to come out of his home infrastructure. This would impact the reasoning on the deployment: the hardware entities would have to be filtered with respect to this new data privacy policy. Moreover, a specific and evolutive ontology for smart cities should be developed to describe the hardware and software component, using the proposed formalism.

4.6. Bibliography

[AIM 06] AIMEUR E., BRASSARD G., FERNANDEZ J.M. *et al.*, "Privacy-preserving Demographic Filtering", *Proceedings of the 2006 ACM Symposium on Applied Computing, SAC'06*, New York, USA, ACM, pp. 872–878, 2006.

[ALB 10] ALBEROLA J., SUCH J., GARCIA-FORNES A. *et al.*, "A performance evaluation of three multiagent platforms", *Artificial Intelligence Review*, Springer, Netherlands, vol. 34, no. 2, pp. 145–176, 2010.

[ARN 08] ARNOLD W., EILAM T., KALANTAR M. *et al.*, "Automatic Realization of SOA Deployment Patterns in Distributed Environments", in BOUGUETTAYA A., KRUEGER I., MARGARIA T. (eds), *Service-Oriented Computing – ICSOC 2008*, vol. 5364 of *Lecture Notes in Computer Science*, Springer Berlin, Heidelberg, pp. 162–179, 2008.

[ATZ 10] ATZORI L., IERA A., MORABITO G., "The Internet of Things: a survey", *Computer Networks*, vol. 54, no. 15, pp. 2787–2805, 2010.

[BAB 15] BABAI L., "Graph Isomorphism in Quasipolynomial Time", *CoRR*, vol. abs/1512.03547, 2015.

[BAR 06] BARTH A., DATTA A., MITCHELL J. *et al.*, "Privacy and contextual integrity: framework and applications", *Security and Privacy, 2006 IEEE Symposium on*, pp. 15–198, May 2006.

[BRA 05a] BRAUBACH L., POKAHR A., BADE D. *et al.*, "Deployment of Distributed Multi-agent Systems", in GLEIZES M.-P., OMICINI A., ZAMBONELLI F. (eds), *Engineering Societies in the Agents World V*, vol. 3451 of *Lecture Notes in Computer Science*, Springer Berlin, Heidelberg, pp. 261–276, 2005.

[BRA 05b] BRAUBACH L., POKAHR A., MOLDT D. *et al.*, "Goal Representation for BDI Agent Systems", in BORDINI R., DASTANI M., DIX J. (eds), *Programming Multi-Agent Systems*, vol. 3346 of *LNCS*, Springer, Berlin Heidelberg, pp. 44–65, 2005.

[CAV 14] CAVAL C., EL FALLAH SEGHROUCHNI A., TAILLIBERT P., "Keeping a Clear Separation between Goals and Plans", in DALPIAZ F., DIX J., VAN RIEMSDIJK M., (eds), *Engineering Multi-agent Systems*, vol. 8758 of *Lecture Notes in Computer Science*, Springer International Publishing, pp. 15–39, 2014.

[CHE 04] CHEN H., FININ T.W., JOSHI A. *et al.*, "Intelligent Agents Meet the Semantic Web in Smart Spaces", *IEEE Internet Computing*, vol. 8, no. 6, pp. 69–79, 2004.

[CHE 05] CHEONG C., WINIKOFF M., "Hermes: designing goal-oriented agent interactions", *Agent-Oriented Software Engineering, 6th International Workshop*, AOSE 2005, utrecht, The Netherlands, 2005.

[CHE 08] CHEIN M., MUGNIER M.-L., *Graph-based Knowledge Representation: Computational Foundations of Conceptual Graphs*, Springer, London, 2008.

[CIS 07] CISSÉE R., ALBAYRAK S., "An Agent-based Approach for Privacy-preserving Recommender Systems", *Proceedings of the 6th International Joint Conference on Autonomous Agents and Multiagent Systems*, AAMAS '07, New York, USA, ACM, vol. 8, pp. 182:1–182, 2007.

[CON 04] CONTE D., FOGGIA P., SANSONE C. *et al.*, "Thirty Years Of Graph Matching In Pattern Recognition.", *IJPRAI*, vol. 18, no. 3, pp. 265-298, 2004.

[COR 01] CORDELLA L.P., FOGGIA P., SANSONE C. *et al.*, "An improved algorithm for matching large graphs", *3rd IAPR-TC15 Workshop on Graph-based Representations in Pattern Recognition, Cuen*, pp. 149–159, 2001.

[CRE 09] CRÉPIN L., DEMAZEAU Y., BOISSIER O. *et al.*, "Sensitive Data Transaction in Hippocratic Multi-agent Systems", *Engineering Societies in the Agents World IX: Lecture Notes in Computer Science*, Springer Berlin/Heidelberg, pp. 85–101, 2009.

[DEA 04] DEAN J., GHEMAWAT S., "MapReduce: Simplified Data Processing on Large Clusters", *OSDI'04: Sixth Symposium on Operating System Design and Implentation*, San Fransisco, 2004.

[DUC 01] DUCATEL K., BOGDANOWICZ M., SCAPOLO F. *et al.*, "Scenarios for ambient intelligence in 2010", 2001.

[EDW 04] EDWARDS G., DENG G., SCHMIDT D.C. *et al.*, "Model-driven configuration and deployment of component middleware publish/subscribe services", *Lecture Notes in Computer Science*, vol. 3286, pp. 337–360, 2004.

[ELF 10] EL FALLAH-SEGHROUCHNI A., OLARU A., NGUYEN N.T.T. *et al.*, "Ao Dai: agent Oriented Design for Ambient Intelligence", in DESAI N., LIU A., WINIKOFF M. (eds), *PRIMA*, *Lecture Notes in Computer Science*, vol. 7057, Springer, pp. 259–269, 2010.

[EVA 11] EVANS D., The Internet of Things: how the next evolution of the Internet is changing everything, White Paper, Cisco, 2011.

[FLI 08] FLISSI A., DUBUS J., DOLET N. *et al.*, "Deploying on the Grid with DeployWare", *Cluster Computing and the Grid, 2008. CCGRID '08, 8th IEEE International Symposium on*, pp. 177–184, May 2008.

[FOX 81] FOX M.S., "An Organizational View of Distributed Systems", *IEEE Transactions on Systems, Man and Cybernetics*, vol. 11, no. 1, pp. 70–80, 1981.

[GEO 10] GEORGANTAS N., ISSARNY V., BEN MOKHTAR S. *et al.*, "Middleware Architecture for Ambient Intelligence in the Networked Home", *Handbook of Ambient Intelligence and Smart Environments*, pp. 1139–1169, Springer, 2010.

[HEL 04] HELLENSCHMIDT M., KIRSTE T., "A Generic Topology for Ambient Intelligence.", in MARKOPOULOS P., EGGEN B., AARTS E.H.L. (eds), *EUSAI*, vol. 3295 of *Lecture Notes in Computer Science*, Springer, pp. 112–123, 2004.

[HOR 04] HORLING B., LESSER V., "A Survey of Multi-agent Organizational Paradigms", *Knowl. Eng. Rev.*, Cambridge University Press, vol. 19, no. 4, pp. 281–316, December 2004.

[ITU 12] INTERNATIONAL TELECOMMUNICATIONS UNION, Overview of the Internet of things, Recommendation Y. 2060, 2012.

[JOH 02] JOHANSON B., FOX A., WINOGRAD T., "The Interactive Workspaces Project: Experiences with Ubiquitous Computing Rooms", *Pervasive Computing Magazine*, vol. 1, no. 2, 2002.

[KOE 67] KOESTLER A., *The Ghost in the Machine*, Hutchinson, 1967.

[KRA 15] KRAVARI K., BASSILIADES N., "A Survey of Agent Platforms", *Journal of Artificial Societies and Social Simulation*, vol. 18, no. 1, p. 11, 2015.

[KRU 10] KRUPA Y., VERCOUTER L., "Contextual Integrity and Privacy Enforcing Norms for Virtual Communities", in BOISSIER O., EL FALLAH SEGHROUCHNI A., HASSAS S. (eds), *MALLOW, CEUR Workshop Proceedings*, vol. 627, 2010.

[LAR 04] LARROSA J., VALIENTE G., "Constraint satisfaction algorithms for graph pattern matching", *Mathematical Structures in Computer Science*, vol. 12, pp. 403–422, 2004.

[MAC 71] MACLANE S., *Categories for the Working Mathematician*, Springer-Verlag, New York, Graduate Texts in Mathematics, vol. 5, 1971.

[MAR 11] MARZ N., Storm: Distributed and fault-tolerant realtime computation, available at: cloud.berkeley.edu/data/storm-beteley.pdf, 2011.

[MCK 14] MCKAY B.D., PIPERNO A., "Practical graph isomorphism, II", *Journal of Symbolic Computation*, vol. 60, pp. 94–112, Elsevier, 2014.

[MEN 02] MENCZER F., STREET W., VISHWAKARMA N. *et al.*, "IntelliShopper: A Proactive, Personal, Private Shopping Assistant", *Proc. 1st ACM Int. Joint Conf. on Autonomous Agents and Multiagent Systems (AAMAS)*, 2002.

[MES 00] MESSMER B.T., BUNKE H., "Efficient Subgraph Isomorphism Detection: a Decomposition Approach", *IEEE Trans. on Knowl. and Data Eng.*, vol. 12, no. 2, March 2000.

[O'H 12] O'HARE G.M.P., COLLIER R., DRAGONE M. *et al.*, "Embedding Agents within Ambient Intelligent Applications.", BOSSE T. (ed), *Agents and Ambient Intelligence*, vol. 12 of *Ambient Intelligence and Smart Environments*, pp. 119–133, IOS Press, 2012.

[RAM 04] RAMCHURN S.D., HUYNH D., JENNINGS N.R., "Trust in Multi-agent Systems", *Knowl. Eng. Rev.*, Cambridge University Press, vol. 19, no. 1, pp. 1–25, March 2004.

[RIC 03] RICCI A., "Agents and Coordination Artifacts for Feature Engineering.", RYAN M.D., MEYER J.-J.C., EHRICH H.-D. (eds), *Objects, Agents and Features*, vol. 2975 of *Lecture Notes in Computer Science*, Springer, pp. 209–226, 2003.

[RIC 11] RICCI A., PIUNTI M., VIROLI M., "Environment programming in multi-agent systems: an artifact-based perspective", *Autonomous and Multi-agent Systems*, vol. 23, no. 2, pp. 158–192, 2011.

[STA 13] STAVROPOULOS T.G., GOTTIS K., VRAKAS D. *et al.*, "aWESoME: A Web service middleware for ambient intelligence", *Expert Systems with Applications*, vol. 40, no. 11, pp. 4380–4392, 2013.

[SUC 12] SUCH J.M., ESPINOSA A., GARCíA-FORNES A. *et al.*, "Self-disclosure Decision-making Based on Intimacy and Privacy", *Inf. Sci.*, Elsevier Science Inc., vol. 211, pp. 93–111, November 2012.

[SUC 14] SUCH J.M., ESPINOSA A., GARCÃA-FORNES A., "A survey of privacy in multi-agent systems", *The Knowledge Engineering Review*, vol. 29, pp. 314–344, 2014.

[TEN 06] TENTORI M., FAVELA J., RODRIGUEZ M.D., "Privacy-aware Autonomous Agents for Pervasive Healthcare", *IEEE Intelligent Systems*, IEEE Educational Activities Department, vol. 21, no. 6, pp. 55–62, November 2006.

[UDU 10] UDUPI Y.B., SINGH M.P., "Information sharing among autonomous agents in referral networks", *Agents and Peer-to-Peer Computing: 6th International Workshop, APZPC 2007*, Springer-Verlag, pp. 13–26, 2010.

[ULL 76] ULLMANN J.R., "An Algorithm for Subgraph Isomorphism.", *J. ACM*, vol. 23, no. 1, pp. 31–42, 1976.

[WEI 93] WEISER M., "Some Computer Science Problems in Ubiquitous Computing", *CACM*, vol. 36, no. 7, pp. 74–84, 1993.

[WES 67] WESTIN A.F., *Privacy and Freedom*, Atheneum, New York, 1967.

[WHI 12] WHITE T., *Hadoop, the definitive guide*, O'Reilly, 2012.

5

ClouT: Cloud of Things for Empowering Citizen's Clout in Smart Cities

ClouT is a joint European–Japanese ICT project for designing, implementing and validating a reference IoT + Cloud architecture for a smart-city ecosystem. This chapter presents an overview of the ClouT concept, the project, and the ClouT reference architecture for smart cities.

5.1. Objective of the ClouT project

The world is facing several challenges that must be dealt with in the coming years, such as efficient energy management, need for economic growth, security and quality of life of its inhabitants. The increasing concentration of the world population living in urban areas puts cities at the center of these challenges. Indeed, on 2% of the Earth's surface, cities use 75% of the world resources [VON 05]. Currently, more than half of the world population lives in cities, and the urban areas of the world are expected to absorb all the projected population growth over the next four decades, at the same time drawing in some of the rural population. These facts make cities crucial actors for the world's sustainable development strategy.

One immediate action of governments in Europe and Japan has been to take measures to transform cities into "smart cities" that better manage their resources. ICT has substantial potential to help cities respond to the growing demands of a more efficient, sustainable and increased quality of life, thus making them

Chapter written by Kenji TEI, Levent GÜREEN and Takuro YONEZAWA.

"smarter". Smartness is directly proportional to "awareness", which is defined in the literature as the state or ability to perceive, to feel, or to be conscious of events, objects, or sensory patterns. With the spectacular advances in sensor and wireless technologies, we are now able to sense the physical events that occur in the environment, pre-process the events with embedded computing capabilities on the sensing devices and transmit them wirelessly. Thus, an important step has been taken toward building the city nervous system that would provide awareness to the city: the Smart City Ecosystem.

To tackle this issue, two recent technologies from ICT have great potential to provide the necessary enablers: the Internet of Things and Cloud Computing.

– *Internet of Things (IoT)* is the set of technologies that can interconnect virtually anything, from daily life objects to more sophisticated networked devices. IoT devices can capture not only the environmental context (temperature, air quality, presence, movement etc.), but also the context of users (preferences, friends, intentions etc.). Some of them may be actuator devices, i.e. acting to change the environment, then giving the power to users of controlling it in accordance with his/her desire. IoT democratizes the way users interact with information, often in real time and related to the physical environment. Efficient integration of this information into existing processes, from business processes to citizens' daily tasks, may have a great potential to increase the smartness of the city.

– *Cloud Computing* provides a flexible virtual execution environment for processing any application over a potentially infinite number of resources, scaling up and down according to usage behaviors. Its decentralized nature enables high reliability and accessibility to information from anywhere at any time. Cloud centers can provide an on-demand metered services at different granularities with a requested quality of service level. Besides new technological enablers, the Cloud paradigm brings new economic models based on pay-per-use that reduce initial investments and related operational costs to the actual use of service, the cost and time for new service providers, in particular for SMEs, enabling entry into a wide market with minimum infrastructure management requirements. They can thus experiment with novel and innovative services whilst reducing the risk of wasting resources.

ClouT is a collaborative project jointly funded by the 7th Framework Program of the European Commission and by the National Institute of Information and Communications Technology of Japan. The ClouT consortium is a strong partnership of leading European and Japanese industries as well as universities and research centers. Table 5.1 shows the consortium of the ClouT project.

	European side	Japanese side
Research centers	CEA-LETI	National Institute of Informatics
University	University of Cantabria	Keio University
Cities	Santander, Genova	Mitaka, Fujisawa
IoT provider	ST Microelectronics	Panasonic
Service/Cloud provider	Engineering	NTT East, NTT R&D

Table 5.1. *Consortium of the ClouT project*

ClouT's overall concept is to leverage Cloud Computing as an enabler to bridge the Internet of Things with Internet of People via Internet of Services. ClouT aims to establish an efficient communication and collaboration platform exploiting all possible information sources, to make cities smarter and help them to face the emerging challenges of efficient energy management, economic growth and development. ClouT will provide infrastructures, services, tools and applications to be reused by different city stakeholders such as municipalities, citizens, service developers and application integrators, to create, use and manage user-centric applications, taking advantage of the latest advances in Internet of Things and Cloud domains. ClouT, with its user-centric approach, will also offer to end-users the possibility of creating their own Cloud services and share them with other citizens.

5.2. Goal of the ClouT project

ClouT's main goal is to design, implement and validate a reference IoT + Cloud architecture for a smart-city ecosystem, that will help city authorities to provide the backbone for the innovation of their environment. ClouT will use Cloud models and technologies as a foundation for bridging the IoT with Internet of People, via Internet of Services, empowering citizens in a ClouT-enabled Smart City.

ClouT will have the following major outputs:

1) a smart-city infrastructure with a near infinite processing and storage capacity for data from trillions of things and people, integrated via virtual services in the Cloud, while keeping their universal interoperability;

2) a set of platform level tools and services aiming at facilitating IoT application development, deployment and supervision;

3) secure data access and processing mechanisms that can handle big data acquired from heterogeneous sources in quasi-real time;

4) innovative city applications and field trials in four pilot cities: Santander and Genova in Europe, and Mitaka and Fujisawa in Japan.

In the ClouT project, we designed a ClouT reference architecture. The ClouT reference architecture shows a well-defined way to develop a smart-city platform, providing functions required by typical use cases of smart-city applications. It will reduce costs and time to develop and use new applications by using new ideas of public–private partnerships inspired by Cloud models, supporting win–win strategies for all stakeholders. ClouT's ultimate goal is to join forces and create a long-lasting synergy for smart-city initiatives between Europe and Japan.

5.3. ClouT concept

Before describing the ClouT reference architecture, we first introduce the ClouT concept. We show a domain model explaining key concepts behind the ClouT reference architecture.

Our approach is to apply the three-layer architecture used in the Cloud domain, which consists of the Infrastructure as a Service (IaaS) layer, Platform as a Service layer (PaaS), and Software as a Service (SaaS) layer. The three-layer architecture allows many applications to use computing resources as needed. We apply this concept to smart cities to enable sharing of various resources in the cities, such as sensors, actuators, data, storage and citizens themselves, among many smart city applications.

Figure 5.1 illustrates an overview of the ClouT domain model. ClouT's overall concept to leverage Cloud Computing as an enabler to bridge The Internet of Things with Internet of People via Internet of Services. Therefore, ClouT provides three-layer concepts inspired by the Cloud domain:

– *City Infrastructure as a Service (CIaaS) layer*:

- provides virtualized city resources (sensor, actuators, storage etc.) as a service;

– *City Platform as a Service (CPaaS) layer*:

- provides a city computing platform (event processing, mash-up tools etc.) as a service;

– *City Software as a Service (CSaaS) layer*:

- provides city application software (city event analyzer, public space management etc.) as a service.

All layers provide their functionality as a service to be shared among users or other software systems.

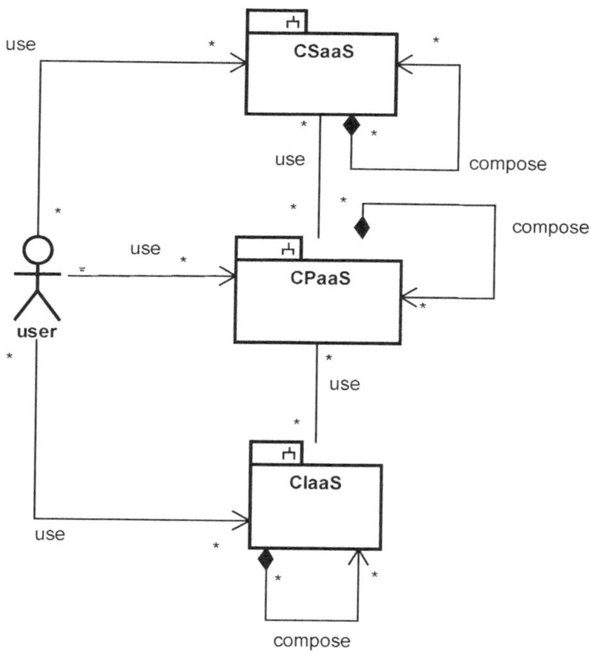

Figure 5.1. *ClouT domain model – an overview*

Figure 5.2 shows a complete view of the domain model. ClouT adopts a representational state transfer (REST) style concept. A service provides a uniform interface to resources, by exposing resources for accessing states of the entities. The style is used across all layers.

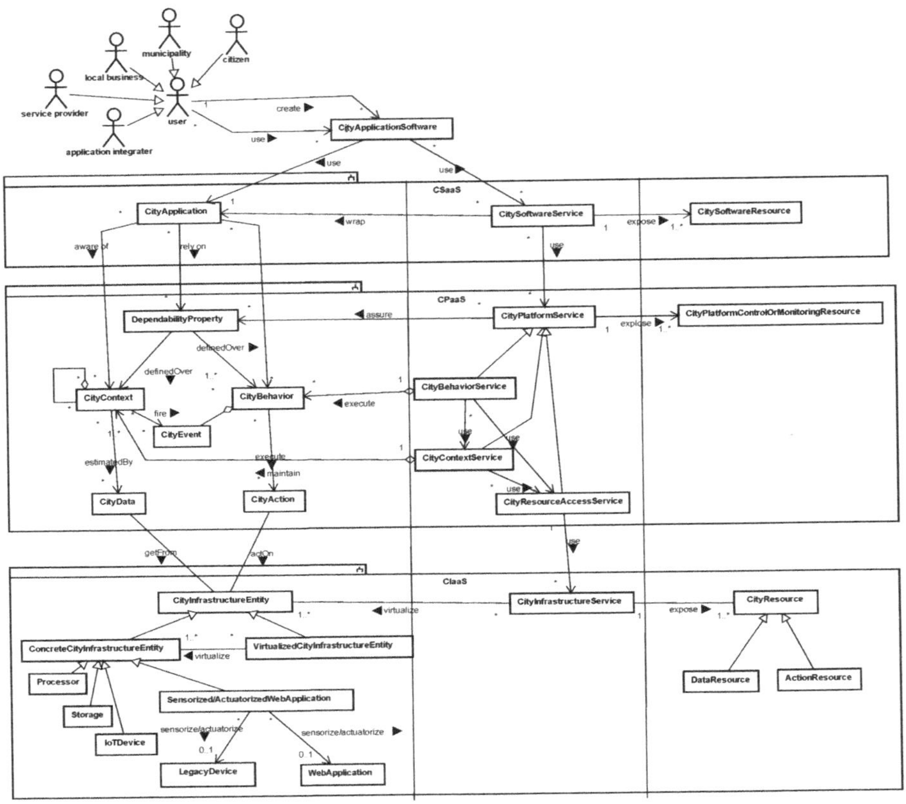

Figure 5.2. *ClouT domain model – a complete view*

5.3.1. *CIaaS concept*

Figure 5.3 shows the details of CIaaS in the domain model. The CIaaS layer provides *CityInfrastructureService* for accessing entities managed in the city infrastructure. Although IaaS in the Cloud domain deals with only computation and storage, CIaaS deals with entities relevant to IoT (such as IoT Devices) and Cloud (such as Processors and Storages). CIaaS also manages Web Applications (such as Twitter or Facebook) and Legacy Devices such as *CityInfrastuructureEntity* by *Sensorization* or *Actuatorization*. The *CityInfrastructureService* provides a uniform interface to *CityInfrastructureEntity* by exposing *CityResources*, in accordance with the REST concept. *CityResource* can be either an *ActionResource*, which manipulates devices, or a *DataResource*, which represents data provided by devices.

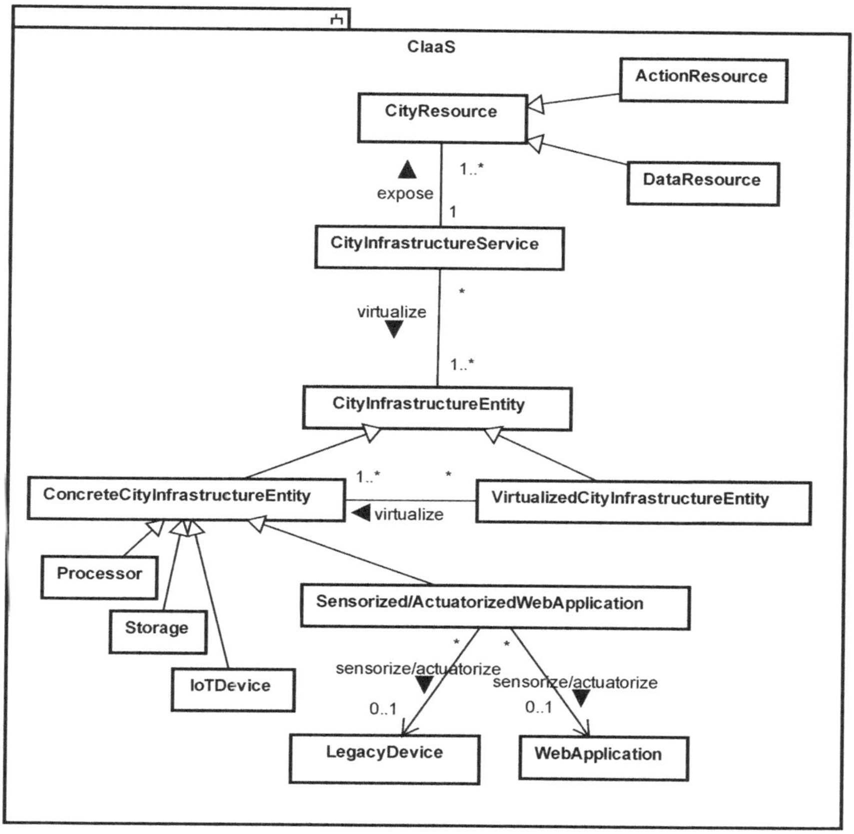

Figure 5.3. *ClouT domain model: ClaaS*

For example, Figure 5.4 shows a Light Service, to manage a Light Device used in a city. The Light Device is a CityInfrastructureEntity and exports the Light Service, which is a *CityInfrastructureService*. The light service exposes "*On*", "*Alert*" and "*Brightness*" resources. "*On*" and "*Alert*" resources (*ActionResource*) represent the state of actions provided by the light devices, and the "*Brightness*" resource (*DataResource*) represents the current brightness of the light devices.

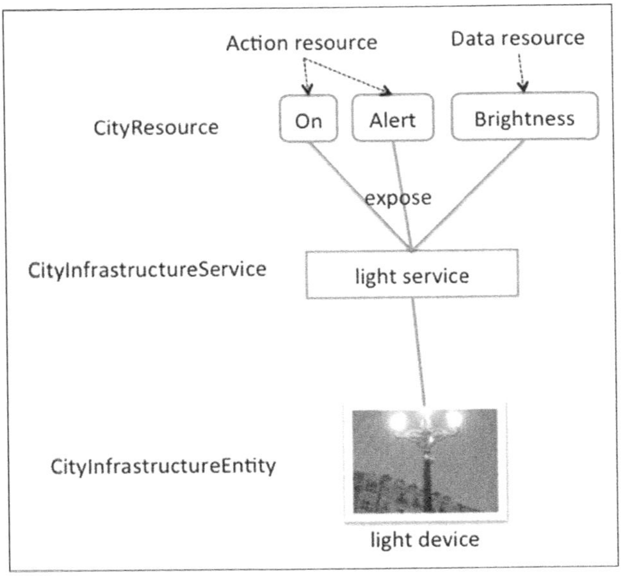

Figure 5.4. *Example of City Infrastructure Service*

ClouT introduces virtualization concepts used in the Cloud domain. *IaaS* in the Cloud domain enables a service to use a virtualized entity (e.g. virtual machine) instead of a concrete entity (e.g. physical machine) to manage resources in a flexible way, by mapping virtualized resources to concrete resources at runtime. *CIaaS* in the ClouT domain supports virtualization of sensors and actuators to improve flexibility and portability. *CIaaS* introduces *VirtualizedCityInfrastructureEntit*ies, representing virtualization of *ConcreteCityInfrastructureEntity*. *CIaaS* enables *CityInfrastructure Services* to use *VistualizedCityInfrastructureEntity* instead of *CityInfrastructure Entity*.

VirtualizedCityInfrastructureEntity is mapped to *ConcreteCityInfrastructure Entity* at runtime. Figure 5.5 illustrates an example of *VirtualizedCityInftastructure Entity*. The *CityLightService* provides an abstract interface to control and monitor some adequate lights in the street. It manages *LightsInStreet* (*VirtualizedCity InstrastructureEnitity*), which does not correspond to a certain light device but is mapped to *ConcreteCityInfrastructureEntity*, like *Light1* or *Light2* at runtime.

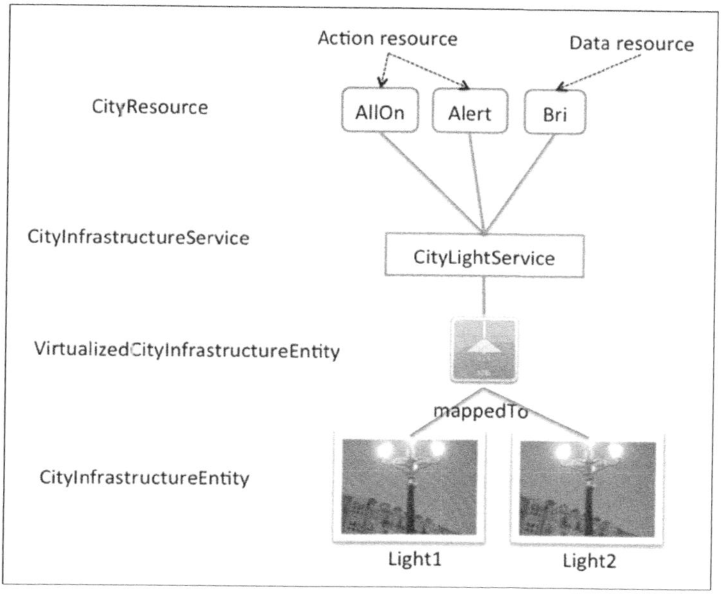

Figure 5.5. *Example of Virtualized City Infrastructure Entity*

5.3.2. *CPaaS concept*

Figure 5.6 shows details of CPaaS in the domain model. The CPaaS layer provides platform services utilizing one or more CIaaS services. We call services provided in the CPaaS *CityPlatformServices*. CIaaS provides three kinds of *CityPlatform Services: CityResourceAccessService*, *CityContextService* and *CityBehaviorService*.

CityResouceAccessService enables access to city resources provided by CIaaS. It provides rich APIs for accessing one or more *CityInfrastructureServices*, which provide primitive APIs for accessing city resources.

CityContextService generates and manages *CityContext* representing the current state of a city, which is determined by analyzing city data obtained from CIaaS. *CityContext* represents the current context of a city, and it is derived from *CityData*. *CityContextService* provides not only APIs for managing *CityContext*, but also event processing APIs to generate *CityContext* from *CityData*. *CPaaS* also supports *CityBehaviorServices*.

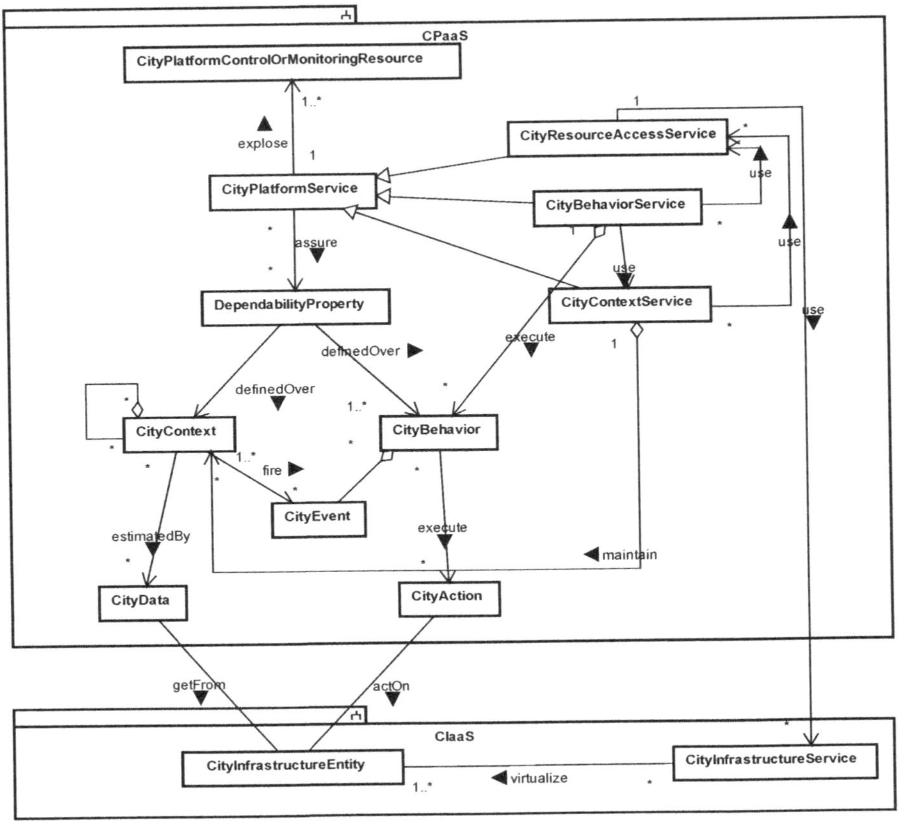

Figure 5.6. *ClouT domain model: CPaaS*

CityBehaviorService supports application developers in composing services provided by ClaaS. The application logic is referred to as *CityBehaviors*. *CityBehavior* can change or can be ruled by *CityContext* and consists of *CityActions* and *CityEvents*. For example, "the street is bright" or "the street is dark" (*CityContext*) are estimated from brightness data (*CityData*) provided by ClaaS. A "sunset" event (*CityEvent*) is initiated by changes in the *CityContext* (from "the street is bright" to "the street is dark"). The service executes *CityActions*, acting on some *CityInfrastructure Entity* corresponding to lights in the street switching on. In addition, CPaaS supports the assurance of *DependabilityProperties* related to *CityBehavior* and

CityContext. *DependabilityProperties* includes a variety of properties to satisfy non-functional requirements, such as secure accessibility to *CityContext* that includes some personal information, or the safety of *CityBehavior* that controls traffic.

5.3.3. *CSaaS concept*

Figure 5.7 shows the details of the CSaaS layer in the domain model. CSaaS allows users to access a city application as a service. The CSaaS layer enables users to build their *CityApplications* by using CPaaS. Each CSaaS may be constructed and operated by a using specific part of the CPaaS functionalities, depending on the application characteristics. For example, data-centric CSaaS should make intensive use of *CityContextService* to realize data processing (analysis, mining, learning etc.). The other CSaaS may require more support on the control behavior to orchestrate multiple services by following interaction protocols.

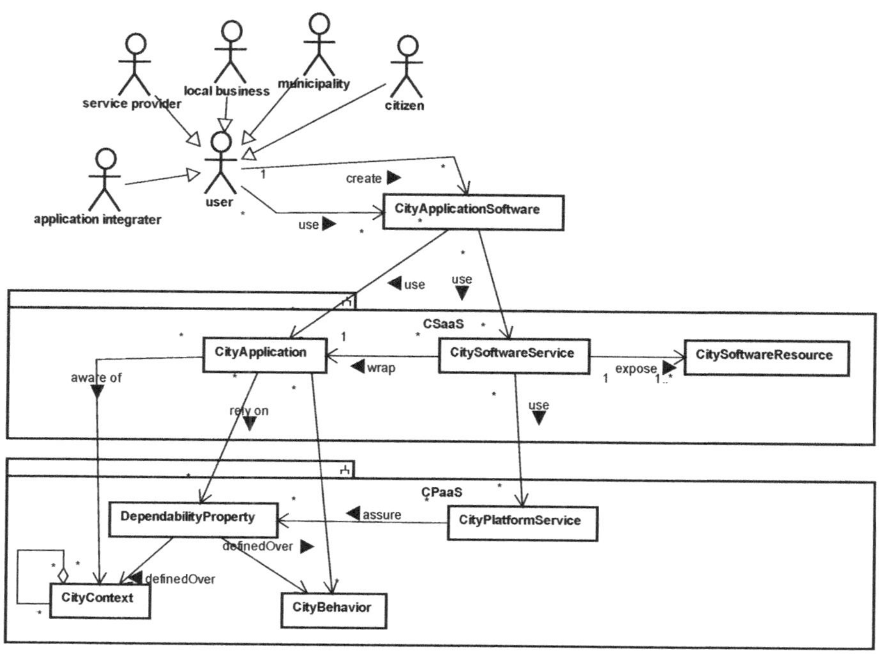

Figure 5.7. *ClouT domain model: CSaaS*

5.4. ClouT reference architecture

In accordance with the ClouT concept, we constructed a ClouT reference architecture. The ClouT reference architecture has been conceived starting from the challenges proposed by four pilot cities: Mitaka and Fujisawa in Japan, and Santander and Genova in Europe. The vision is Cloud-centric, i.e. Cloud service models (IaaS, PaaS and IaaS) represent the core supporting the modules that bind together Internet of Things, Cloud Computing and Internet of People. This renders municipalities capable of using scalable infrastructures that allow the gathering, processing and exploitation of huge amounts of data harvested from physical devices, the Web and participatory sensing. ClouT's architecture includes both easy-to-use and fully-fledged scalable development platforms that cater to citizens and developers, similar to enabling the development of Cloud-enabled smart city services. Figure 5.8 shows a high-level view of the ClouT reference architecture, detailing the composing layers, functional blocks and models.

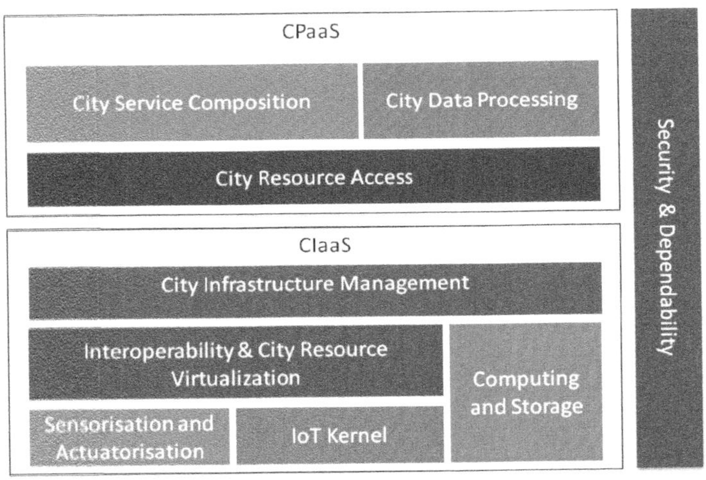

Figure 5.8. *ClouT reference architecture overview*

5.4.1. *CIaaS components*

The CIaaS layer provides the required facilities to exploit all the computing and sensing resources available to the City, while making it easier to extend such infrastructures through the adoption of Cloud technologies, standard protocols and interoperability facilities. Figure 5.9 shows the logical representation of the main functional blocks within the CIaaS layer.

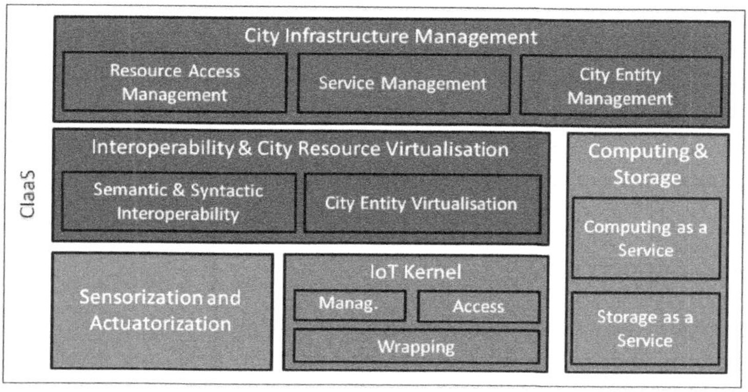

Figure 5.9. *Main functional blocks for City Infrastructure layer*

Here is a brief breakdown of its main subcomponents:

– *City Infrastructure Management* offers centralized search capabilities and event management for discovered city resources. It enables tracking of all the available city resources, detecting changes in state and offers information on the availability of devices and services.

– *Computing and Storage* offers all the physical and virtualized computing, networking and storage resources that are required to store, retrieve and elaborate city data. It has load balancing, high availability and scalability qualities that are typical of the Cloud. It's the foundation on which the other city service models lay.

– The *Interoperability & City Resource Virtualization* component is in charge of validating and converting data gathered from both IoT and sensorized devices. Its duty is to transform raw, unstructured city data into meaningful data in context. It offers both syntactic and semantic interoperability capabilities. Moreover, it includes sensor/actuator virtualization capabilities, offering extended sensor/actuator abstraction functionalities.

– The *Sensorization and Actuatorization* component is responsible for the "sensorization" of legacy devices, rendering such resources IoT-compliant sensors and/or actuators. This sensorization/actuatorization process applies to both physical legacy devices and selected Internet of People sources, such as social networks. This component includes the facilities required to transform legacy resources into smart objects, the noise reduction functionalities required to extract meaningful data from external Web data sources, and offers access to the sensorized resources via APIs.

– The *IoT Kernel* comprises all the IoT resources available to the City. It includes a multiprotocol IoT Gateway that is in charge of gathering heterogeneous IoT device's data streams. It offers a multiprotocol compatibility layer that complies with all IoT transmission protocol standards as well as uniform access to the connected resources. It has discovery and management functionalities that enable the remote management of the bound IoT sensors/actuators.

5.4.2. *CPaaS components*

The CPaaS layer includes the development and processing tools offered by the ClouT platform. It lays the foundations on which the end-user services are built. Figure 5.10 displays the logical representation of the main functional blocks within the City Platform layer:

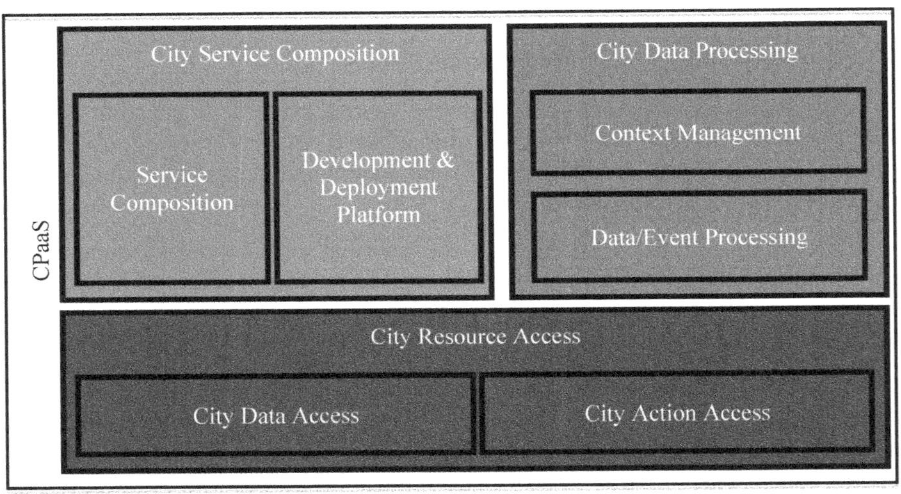

Figure 5.10. *Main functional blocks for the City Platform layer*

Here is a brief breakdown of its main subcomponents:

– *The City Service Composition* block aims at both technically oriented and un-experienced users, to mash-up and aggregate data and services offered by used applications. *The Service Composition component* has a graphical user interface to enable accessible interaction with the composition services. *The Development and Deployment Platform component* caters to the experienced developer with platform as a Service Cloud capabilities, enabling the swift realization of customized development environments and the shipping of the developed applications on auto-scalable, self-balancing virtualized instances.

– *The City Data Processing* block aims to build the data processing layer of the ClouT platform. It gathers city data from the city resource access component and provides processed data/events, as well as adding high-level contextual information to the applications or to the service composition environment. *The Data/event processing* component is in charge of handling (storing and processing) data collected by various city data and event sources. *Context management* is in charge of storing and delivering high-level context information obtained from the processed data and events.

– *City Resource Access* offers the middleware software that enables the storing and retrieval of the collected city data, together with the associated metadata, from and into the backend storage via a uniform data access layer. It's a gateway to the City IaaS layer.

5.4.3. *Security and Dependability components*

The Security and Dependability block gathers all the required security functionalities, included the protocols used, needed to check and authorize the access to the all the modules developed and used, including CSaaS. In particular, CSaaS applications can leverage the Security and Dependability Module or use their own A.A.A. system.

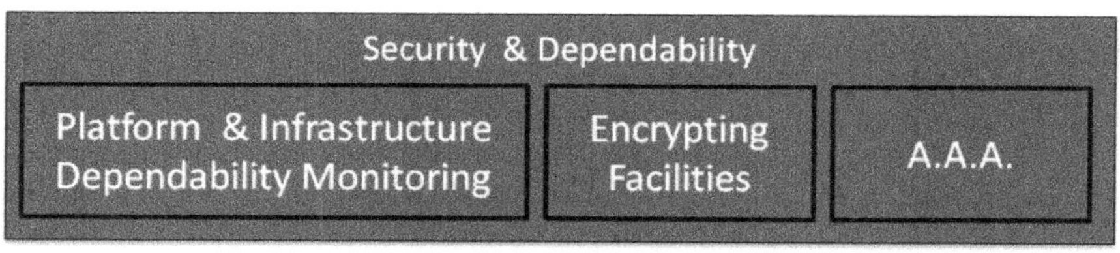

Figure 5.11. *Functional Components for Security & Dependability*

Here is a brief breakdown of its main subcomponents:

– *The City Authentication, Authorization and Accounting (A.A.A)* component offers authentication and authorization facilities to the CIaaS and CPaaS layers. This component has scalability mechanisms to manage a great number of account and authorization rules. It exposes standard API-based access to the City Infrastructure as a Service and City Platform as a Service layer. By ClouT architecture design choice, City Applications (within CSaaS) can leverage the ClouT security framework or use their own internal security modules; to do so, such Applications must be trusted by the ClouT infrastructure.

– *The Encrypting Facilities* component encompasses all the technologies and protocols used to encrypt stored data and communications between the applications and the other components running in the platform on each layer of the architecture (CSaaS, CPaaS and CIaaS). Typically this includes technologies such as SSL and TLS with various encryption algorithms.

– *The Platform & Infrastructure Dependability Monitoring* component monitors all the resources, hardware and software running on CIaaS and CPaaS layers. To monitor the available resources, both passive – remote probing of status – and/or active – an agent local to each device/service – may be appropriate depending on circumstances. The monitoring facility is capable of checking the status of the monitored devices/service with a minimal footprint and transmission overhead, while being highly customizable.

5.5. Mapping the architecture

This section focuses on the mapping of the ClouT reference architecture into a real-world scenario by application.

A generic real world scenario is composed of different types of actors that interact with the architecture to generate and offer a service (Municipality and Third Party developers), consume a service (citizens and Municipality users) and manage the infrastructure/platform (system administrators and infrastructure providers). For each of these actors, the interaction with the architecture is carried out in the following way:

– First of all, and in a transversal way to CIaaS, CPaaS and CSaaS layers, it utilizes the *Security & Dependability* module, in charge of the authentication and authorization of the users accessing the platform. This translates into the assignment of different access policies and permissions according to the user role. In this sense, as detailed in the Security and Dependability module, some applications in the CSaaS layer will use the authorization and authentication capabilities provided by the ClouT architecture, whereas the others will use their own mechanisms trusted by ClouT architecture.

– *Citizens and Municipality users*: these users access the city applications provided by the CSaaS layer which have been created for using the corresponding tools provided by the CPaaS, using the resources and the data stored and processed at the CIaaS platform. These are the users that only consume information from the platform by using the corresponding applications, but they are neither qualified nor authorized to access the CPaaS and CIaaS layers. In principle, these applications, created by a third-party service provider and application developers, could be part of an other architecture using a different security mechanism trusted by the ClouT architecture.

– *Municipality and Third-Party developers*: these users want to access the architecture to use the resources and services offered by the platform, as well as to create and develop new applications, mainly at the CSaaS layer. For this purpose, as a first step, the users access the architecture through the *Security and Dependability* module located in the CPaaS layer, being assigned the corresponding permissions according to the user role. Once within the platform, the user can access the *City Service Composition* block that provides the corresponding mechanisms to create and develop services as a composition of services from the city. Additionally, data from the CIaaS can be accessed and processed through the *City Resource Access* and *City Data Processing* modules, respectively. The communication between CPaaS and CIaaS will be carried out through the communication between the *City Resource Access* module in CPaaS and the *City Infrastructure Management* block in CIaaS. This module provides the corresponding mechanisms for accessing and managing the sensors, actuators, IoT devices, SNS and other information sources, including both physical and virtual resources which compose the city platform, thus offering these data in a standardized way to the CPaaS layer.

– *System administrators and infrastructure providers*: these actors will mainly access the CSaaS and the CIaaS layers. Regarding the CSaaS layer, applications for managing and monitoring the infrastructure as well as handling events and actuating over the used platform will be accessed by this user type. These applications can be developed by using components in the ClouT platform. With respect to CIaaS layer access, this will be carried out through the *Security and Dependability* module, being assigned the corresponding credentials for accessing the CIaaS modules. In this sense, unlike Municipality and Third-Party developers, these users will directly access the different resources provided by the platform of *IoT Kernel* and *Sensorization and Actuatorization* modules, thus modifying capabilities, and adding or removing the resources that compose the platform. In addition to this, the *Interoperability & City Resource Virtualization* module formats and homogenizes data coming from the lower modules. All the information retrieved by these resources will be processed and stored by the *Computing & Storage* module, offering a scalable and dependable Cloud computing infrastructure for exploiting resources and city services, as well as storing the data provided by these resources. Finally, as previously indicated for Municipality and Third-Party developers, the *City Infrastructure Management* module will act as a bridge between CIaaS and CPaaS, thus managing and offering data from infrastructure resources to the upper layers.

Having defined the interaction of each user type with the ClouT reference architecture, a Traffic Mobility Management use case is taken as an example to clarify these interactions between these three types of users and the three layers composing the ClouT reference architecture.

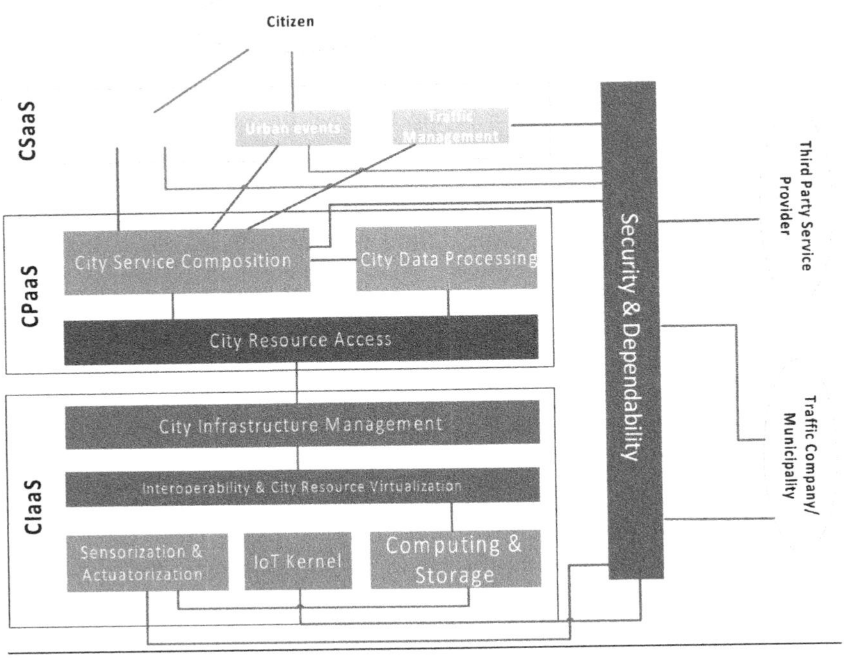

Figure 5.12. *Interaction with ClouT reference architecture in a traffic mobility management use case*

Figure 5.12 displays the interactions among the different users within the ClouT reference architecture:

– The *citizens* access different applications (within the CSaaS) that provide route planning according to certain user requirements and necessities and also to the monitoring of the urban events occurring in the city. As can be observed in the figure, these types of applications could be developed by additional third-party users using their own security mechanism and accessing to the Security and Dependability block as a trusted application.

– The *traffic company* and the *Municipality* will use the traffic management application (in the CSaaS layer) for monitoring and managing the available resources in the lower layer. In this sense, the application can be developed upon the

ClouT platform. Apart from that, they access the CIaaS to add new IoT devices, sensors and actuators for providing the corresponding data for the traffic mobility use case. All these data will be processed and stored in the Computing & Storage block, formatted accordingly in the Interoperability & City Resource Virtualization module, and made available to the City Infrastructure Management block.

– *Third-Party service providers* will enter the platform through the Security and Dependability block, thus accessing to the Service Composition to generate the different applications of the CSaaS layer. In this sense, the City Service Composition will be fed by the data and events processed by the City Data Processing block and will be able to access and manage the different city and services/resources using the City Resource Access.

As exemplified by this Traffic Mobility Management use case, the different users interact with the different layers of the platform, producing and/or consuming the information provided by the subjacent platform. This interaction will be further detailed in the WP4 deliverables (D4.2 and D4.3) [CLO 16], which descnite real implementations and deployments of the corresponding applications to be used in the four trial cities of the project.

5.6. Conclusion

The proposed user scenario shows how the ClouT approach can be applied in a real world context. An important aspect of the context is the management of a relatively high amount of data in quasi–real time. This applies to both a high quantity of data and a burst of data concentrated in a limited time slot. In general, modern cities, especially if willing to offer several high quality services, must face the issues related to the described aspects: for example, a simple traffic jam can provoke sudden access by numbers of drivers to smartphone applications to find out the cause and potential alternative routes. Moreover, special public events such as exhibitions, sporting events or natural disasters can increase the amount of data processed and used for a certain period of time longer than a simple burst. These events are not so unusual in cities, regardless of their size: they can occur in big cities as well as in small cities.

These issues are not the only ones to be considered. The efficiency of public services for young and old people, emergency or healthcare is strictly related to the quantity and the quality of information shared between municipalities and citizens.

For all the use scenarios considered, the data must be collected as finer-grained as possible: sensors managed by gateways provide a continuous flow of information that should be stored, processed together with other data and used to produce other information that, in turn, should be stored. Moreover, time-limited data bursts lasting from few seconds to few days should be processed in quasi–real time, and the implementation of reactions to either simple or very complex alerts must always be supported by the infrastructures of the city.

The simplest solution is to provide the hardware needed to store the required data for an appropriate amount of time and to face the load peaks in the worst case. The costs are enormous, totally unaffordable for small municipalities, but also very difficult to sustain for big cities, especially if a good quality of service is requested.

The ClouT approach defines how Cloud services and IoT can cooperate to obtain significant benefits in offering services for Smart Cities. The ClouT approach demonstrates that it can be specialized for application to SmartCities. In particular, the infrastructure layer (CIaaS) enables the abstraction of the concept of a sensor, also taking into account social networks, which are a very precious data source, and providing an unlimited and secure storage capability. The platform layer (CPaaS) enables the composition of services and data processing, enabling the developers of the software layer (CSaaS) to offer more and more innovative services. Cloud technology provides these features with elasticity and cost-effectiveness, enabling small (but also big) municipalities to provide services on a scale as required, and to pay only for the resources used.

For these reasons, the ClouT approach has the potential to pave the way for small municipalities willing to make a first step into the world of the smart-city easily and cheaply, while making already *smart* cities even *smarter*.

5.7. Bibliography

[CLO 16] CLOUT CONSORTIUM, ClouT Deliverables, available at: http://clout-project.eu/documents/deliverables-access, 2016.

[VON 05] VON UEXKULL J., GIRADET H., COUNCIL W.F., *Shaping Our Future: Creating the World Future Council*, Green Books, available at: http://books.google.co.jp/books?id=NQnXAAAAMAAJ, 2005.

6

sensiNact IoT Platform as a Service

With the spectacular progress in computing, storage, communication and sensor technologies, smart and autonomous devices are being increasingly used and are beginning to bring the Internet of Things (IoT) to reality. The IoT is a global network, service and data infrastructure in which "things" communicate and collaborate. According to recent surveys, we expect to have more than 50 billion connected "things" by 2020, including not only traditional communicating devices, such as smart phones, tablets, PCs, connected machines and others, but also everyday objects such as home appliances, water/gas/electricity meters, vehicles, street lamps, parking lots, wearable health devices, watches, coffee cups, sofas, vineyards, cows and others. Continuous awareness of the physical world in real time brings substantial value to existing business processes in various domains such as smart homes, buildings, offices, factories, transport, health care and cities. Innovative and disruptive applications are expected to emerge and revolutionize our life, as has been the case with the emergence of mobile computing.

The unprecedented number of connected things and the need to process the associated big data naturally raise new technical challenges, such as interoperability of data and connectivity heterogeneity. Currently, a large number of commercial and academic IoT platforms exist at the forefront of research, trying to address those challenges. Many of them are application specific, thus suffer from limited reusability and, as a result, from limited sustainability. Currently, many objects in the IoT world are unable to share data with one another and can only be controlled from a single application or system, making current IoT applications vertical and domain specific. IoT is facing this interoperability challenge and lacks horizontal platforms providing the glue to bind the "things" together.

Chapter written by Levent GÜRGEN, Christophe MUNILLA, Rémi DRUILHE, Etienne GANDRILLE and Jander BOTELHO DO NASCIMENTO.

There have recently been propositions targeting horizontal IoT platforms. Those platforms either focus on physical gateways with direct connection to the devices, or virtual platforms in the cloud, collecting and analyzing the data sent by miniaturized sensor devices. We believe that both local processing as well as processing in the cloud are essential for end-to-end efficient IoT platforms. In addition, the existing platforms fail to provide easy-to-use tools for the rapid development and deployment of IoT services and applications.

This chapter presents a service-oriented IoT platform, sensiNact, which brings together physical devices, gateways and virtualized platforms in the Cloud, thus distributing data processing and intelligence at different layers (aka Fog/edge computing). sensiNact is composed of a set of enablers and services that provides the means for building context-aware applications for smart connected objects. It provides generic APIs to access resources provided by IoT devices. The platform integrates different IoT devices and communication technologies to provide homogeneous access to the underlying heterogeneous networks. The platform comes with a tool, sensiNact Studio, that provides rapidly developing applications for developers, without requiring the knowledge of the underlying IoT technologies.

This chapter will present the general concept of the sensiNact platform and sensiNact Studio, as well as details for their implementation and validation via various IoT applications.

6.1. Introduction

The IoT world is a heterogeneous environment where devices, i.e. sensors and actuators, communicate with different network protocols, and have their own data gathering APIs and data models. This heterogeneity of protocols and data models is a barrier to the development of cross-domain applications that could benefit from the variety of existing IoT devices and functionalities they can provide.

This is especially critical for horizontal domains, such as smart cities, that bring together various other domains such as transport, shopping, health care, entertainment, security, public services and others. New and innovative applications could exploit the countless possibilities that IoT devices can provide. Smart cities are composed of hundreds of devices, making the creation of applications increasingly complex. The event-based and on-demand nature of IoT data should be taken into account by the data platforms. Thus, suitable approaches are required to handle this complexity, and to ease the development of applications to automate the interactions between the devices used in the city.

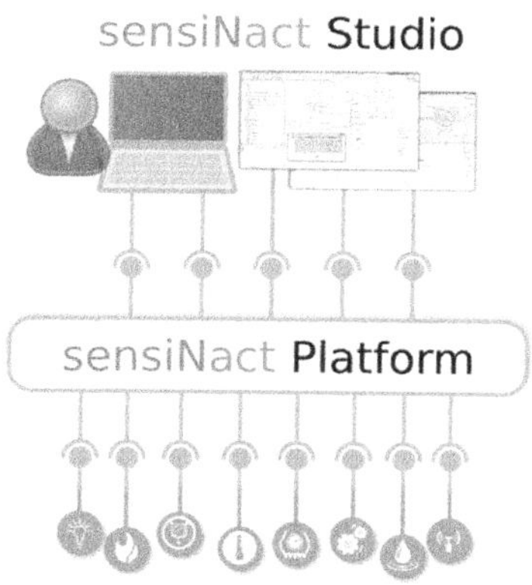

Figure 6.1. *sensiNact platform providing a homogeneous view over heterogeneous IoT devices; sensiNact Studio enables developers to rapidly prototype IoT applications*

In this chapter, we present sensiNact, a horizontal platform dedicated to IoT and used in various smart cities and smart home applications. sensiNact:

– manages IoT protocols and device heterogeneity and provides a unified data model;

– provides synchronous (on demand) and asynchronous (periodic or event based) access to sensor data, as well as access to historic data;

– manages the access rights to devices (sensor and actuator) for third-party users;

– reduces the development time of IoT applications and facilitates their deployment and management at runtime.

To achieve these objectives, sensiNact comes with two complementary frameworks (see Figure 6.1):

– *sensiNact platform* interconnects IoT devices using different southbound IoT protocols, such as Zigbee, EnOcean, LoRa and XBee and allows access to them with various northbound protocols such as HTTP REST, MQTT, XMPP, JSON RPC and CDMI. The platform can also host applications and manage them using an application manager module;

– *sensiNact Studio* proposes an Integrated Development Environment (IDE) based on Eclipse to manage the existing devices, in addition to develop, use and manage IoT applications.

6.2. State of the art

This section details the existing platforms related to sensiNact. First, we detail the architecture of existing IoT solutions. Then, we briefly present each solution and compare them with sensiNact.

6.2.1. *IoT solutions architectures*

To ease the analysis of the IoT solutions, we gather them into four categories (see Figure 6.2).

The "end-to-end solution" aims at quickly releasing an IoT product. This solution provides the required software components to connect your device to the middleware in charge of collecting data. However, developers cannot use commercial off-the-shelf (COTS) or already used devices, i.e. the heterogeneity of protocols is not taken into account.

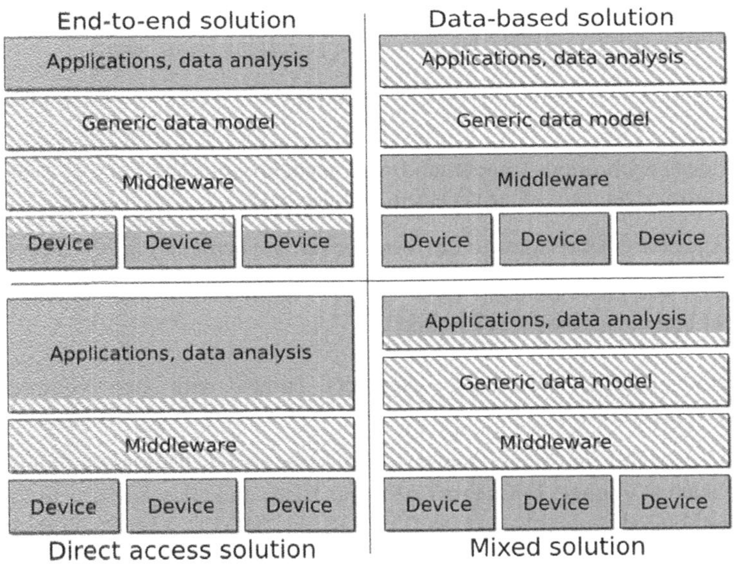

Figure 6.2. *Architecture of IoT solutions. The solid boxes correspond to the complete IoT stack. The solution coverage is displayed in grey hatch*

The "data-based solution" aims at formatting and processing data, enabling the developer to create applications based on it. However, this solution does not provide any mechanism to connect your devices to the layer in charge of the data.

The "direct access solution" only provides the connection layer to the devices and handles heterogeneity of protocols. This solution is a raw solution, in which the developer has direct access to the data. However, this lack of data format may result in difficulties for creating applications.

The "mixed solution" aims at handling protocol heterogeneity at the bottom layer, enables the access to data from devices and simplifies the creation of applications.

6.2.2. Existing IoT platforms

Table 6.1 provides a (non-exhaustive) list of existing IoT solutions and compares them with respect to the following properties:

IoT architecture: See IoT solutions architectures. The choice of the IoT platform architecture influences the remaining work of third-party developers to handle the devices in a given application domain and to build and use applications.

Deployment location specifies whether the solution can be used in the cloud, if it can be used locally (on a local server) or both. Having the possibility to choose where to use the solution gives more freedom to the developers in the construction of their offer.

Application domain details for which domain the solution is designed. Some platforms are designed for a specific environment, others are agnostic to any particular environment, enabling use in other contexts, or in cross-context such as smart cities.

Open source specifies whether or not the IoT solution is open source. The choice of an open-source or of a commercial platform determines the availability and the transparency of the data. An open-source platform enables the developer to better understand the platform and possibly to improve it.

COTS devices integration: indicates whether it is possible to integrate COTS devices. In large scale deployments, such as smart cities, it is unlikely that all the devices come from the same provider. Being able to integrate devices from various providers has to be taken into account for the choice of the IoT platform.

Service oriented: specifies if the IoT solution is service-oriented. Service-oriented approaches, such as OSGi [ALL 12], ease the development of applications and allow easy integration of COTS devices.

As is observed from the table below, the smart home and smart city are the most popular domains. Solutions such as OpenHAB or Hobson, for instance, enable us to connect consumer electronic devices to a local gateway and access them using HTTP requests. For smart cities, the main approach is to use the "data-based solution". This solution, adopted by IBM Bluemix or Exosite, is usually used in the cloud. The owner of the devices has to find a way to send data to the cloud in order to analyze or browse it.

Agnostic environments adopt end-to-end architecture. KAA, Thinger and ThingWorx provide SDK for the devices and a platform to get data from those devices, and enable developers to create an IoT solution from the device to the application layer.

sensiNact is an open-source service-oriented agnostic platform adopting a mixed solution, where application can be used both at the gateway and cloud levels.

The architecture and data model section presents the sensiNact platform, the generic proposed data model and considers heterogeneity in IoT environments. The Platform Security Management section details the security embedded in sensiNact to secure the use of the framework. Finally, the sensiNact Studio section describes the tool developed to visualize the devices and to create applications.

PLATFORM	IOT ARCHITECTURE	DEPLOYMENT LOCATION	APPLICATION DOMAIN	OPEN SOURCE	COTS DEVICES INTEGRATION	SERVICE ORIENTED
ECLIPSE KURA [KUR 16]	Direct access	Cloud, local	Agnostic	✓	✓	✓
EXOSITE [EXO 16]	Data-based	Cloud	Agnostic	×	×	×
HOBSON [HOB 16]	Direct access	Local	Smart home	✓	✓	✓
IBM BLUEMIX [IBM 16]	Data-based	Cloud	Smart city	×	×	×
IOTSENS [IOT 16]	Data-based	Cloud	Smart city	×	×	×
KAA [KAA 16]	End-to-end	Cloud, local	Agnostic	✓	×	✓
OPENHAB [OPE 16b]	Direct access	Local	Smart home	✓	✓	✓
OPEN THE BOX [OPE 16a]	Mixed solution	Local	Smart home	×	✓	✓
THINGER [THI 16a]	End-to-end	Cloud	Agnostic	✓	×	×
THINGWORX [THI 16b]	End-to-end	Cloud	Agnostic	×	×	×
SENSINACT	Mixed solution	Cloud, local	Agnostic	✓	✓	✓

Table 6.1. *Comparison of IoT platforms*

6.3. Architecture and data model

The sensiNact platform allows the interconnection of IoT devices supporting different network protocols, to achieve seamless homogenous access to those embedded devices. The platform is composed of six functional groups and their relative interfaces:

– the *Device Protocol Adapter* abstracts the specific connectivity technology of wireless sensor networks. It is composed of the bridges associated with protocol stacks. All the bridges comply with a generic Device Access API used to interact with northbound sensiNact services;

– the *Smart Object Access and Control* implements the core functionalities of sensiNact such as discovering devices and resources, as well as securing communication among devices and consumers of their services;

– the *Consumer API* is protocol agnostic and exposes services of the Smart Object Access and Control functions to Consumers;

– the *Consumer Protocol Adapter* consists of a set of protocol bridges, translating the Consumer API interface into specific application protocols;

– the *Platform Management* functional group includes all the components needed to ease management of devices connected to sensiNact, regardless of their underlying technologies. A Device Management API is used for this purpose. This functional group also contains the components managing cache, resource directory and security services. These management features are exposed by means of the Platform Management API;

– and finally the *Manager Protocol Adapter* allows adaption of the Platform Management API to the specific protocols used by different additional management entities.

In terms of connectivity (see Figure 6.3), on the southbound side, the sensiNact platform allows us to cope both with "physical device" and "virtual device" protocols, allowing uniform and transparent access to, for instance, an XBee network or an HTTP Restful Web service. A non-exhaustive list of supported protocols is as follows:

– *EnOcean*, concerting energy harvesting wireless sensor technology (ultra-low-power radio technology for free wireless sensors) and protocols in use to interact with those sensors;

– *Bluetooth Low Energy*, which is a personal area network, low-power protocol designed mainly for health care or entertainment applications;

– *MQTT*, which is a machine-to-machine protocol, lightweight publish/subscribe messaging transport, useful for connections with remote locations where a small code footprint is required and/or network bandwidth is at a premium;

– *ZigBee-based protocols* (e.g. XBee);

– *CoAP*, which is a REST application protocol, designed to be "the HTTP for constrained networks and devices" whose concept originated from the idea that "the Internet Protocol could and should be applied even to the smallest devices" and that low-power devices with limited processing capabilities should be able to participate in the IoT; it is usually used with a 6LoWPAN network, but it may also travel between regular IP networks (it is used by the OMA LWM2M protocol, for instance).

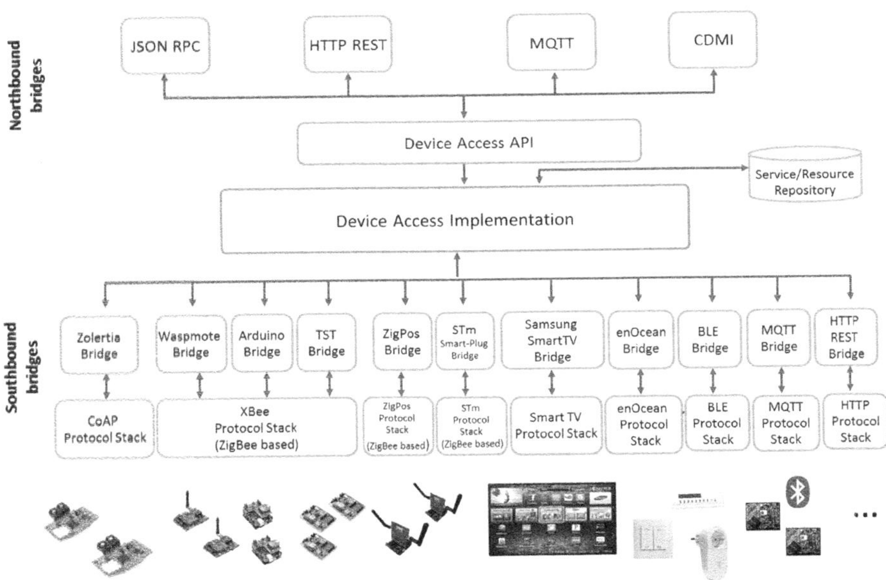

Figure 6.3. *sensiNact platform internal architecture*

On the northbound side the sensiNact platform provides both client/server and publish/subscribe access protocols:

– *MQTT*;

– *JSON-RPC* (1.0 and 2.0);

– *HTTP RESTful*;

– CDMI.

The *Smart Object Access and Control* functional group described above in this section includes a large number of functionalities:

– It handles the communication with the Consumer Protocol Adapter (REST API, JSON RPC etc.) and IoT (and non-IoT) devices, providing URI mapping, incoming data/messages translation in an internal format and outgoing data/messages translation in a Consumer format. Whenever a Consumer tries to access a resource via Consumer API, the requested URI is forwarded to the Resource Manager to check whether or not a specific resource descriptor exists inside the Resource Directory and to verify its accessibility status. If no resource descriptor exists, a message response with an error code is returned to the Consumer API. Otherwise, the request is forwarded to the right interface. At the same time, whenever a response is originated from IoT device (or abstract IoT device), it will also be forwarded to its logical counterpart to update the resource representation in the platform.

– It manages the subscription/notification phases for the consumer, if it is not handled by the targeted device (service) itself.

– It supports Devices and Resource Discovery and Resource Management capabilities, to keep track of IoT Resource descriptions that reflect those resources that are reachable via the platform. These can be both IoT Resources, or resources hosted by legacy devices that are exposed as abstracted IoT Resources. Moreover, resources can be hosted on the platform itself. The Resource Management functionality enables us to publish resources in sensiNact and also for the Consumer to discover what resources are actually available from the platform; sensiNact Service and Resource model allows exposing the resources provided by an individual service. The latter, characterized by a service identifier, represents a concrete physical device or a logical entity not directly bound to any device. Each service exposes resources and could use resources provided by other services. Figure 6.4 depicts the Service and Resource model.

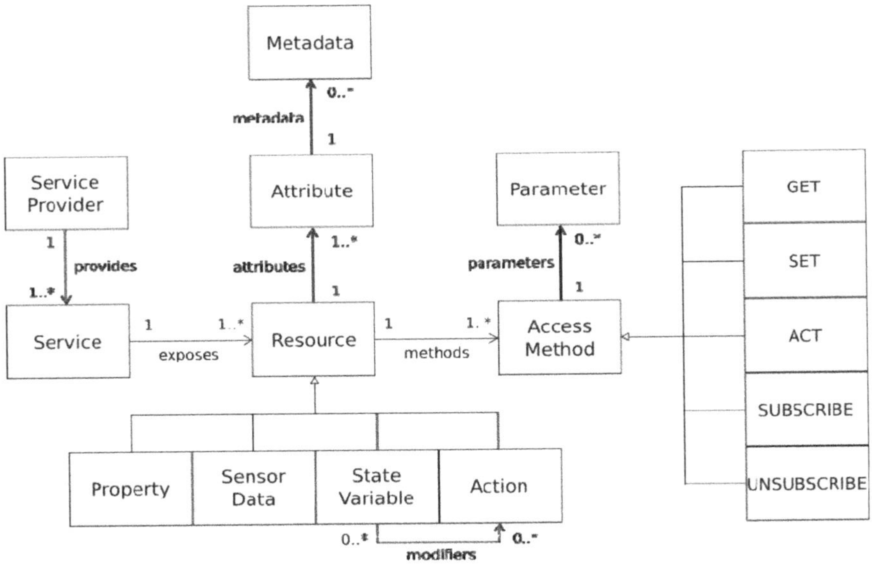

Figure 6.4. *The generic data model of sensiNact allows a uniform access to sensors and actuators*

The *Resource Directory* stores information, i.e. resource descriptions, about the resources provided by individual devices connected to sensiNact. It also supports looking up resource descriptions, as well as publishing, updating and removing resource descriptions to it.

Discovering and using resources exposed by Services is a favored approach to avoid using static service interfaces and then increase interoperability. Therefore, sensiNact Services and their exposed resources are registered into Service/Resource repositories. The platform uses the OSGi service registry as a Service/Resource repository, whereas resources are registered as service properties. Clients ask the Service/Resource repository for resources fulfilling a set of specified properties (defined by LDAP filters). In response, the Service/Resource repository sends clients the list of service references that expose the requested and authorized resources. Clients can then access/manipulate the resources exposed by their selected service objects. Figure 6.5 depicts the typical service-oriented approach, in which when a service provider starts, it registers/publishes itself into the service broker. When a consumer appears, it looks for the reference to the service provider into the service broker. With this reference, the consumer can interact directly with the provider.

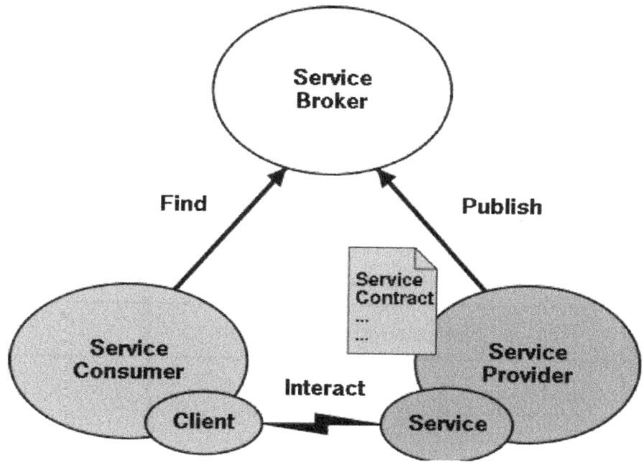

Figure 6.5. *Service-oriented architecture [MAC 06]. Resources and services can be exposed for remote discovery and access using different communication protocols, such as HTTP REST, JSON-RPC and others, and advanced features may also be supported (as semantic-based lookup). Resources can be classified as shown in Table 6.2, whereas the access methods are described in Table 6.3*

Type	Description
SENSORDATA	Sensor data provided by a service (from a physical or virtual device). This is real-time information provided by e.g. the SmartObject that measures physical quantities
ACTION	Action provided by a service. This is mostly an actuation on the physical environment via an actuator associated with the SmartObject supporting this functionality (turn on light, open door, etc.) but can also be a request to do a virtual action (play multimedia on a TV, make a parking space reservation etc.)
STATEVARIABLE	Information representing a SmartObject state variable of the service. This variable is most likely to be modified by an action (turn on light modifies the light state, opening door changes the door state etc.) but also to intrinsic conditions associated with the working procedure of the service
PROPERTY	Property exposed by a service. This is information which is likely to be static (owner, model, vendor, static location etc.). In some cases, this property can be modified

Table 6.2. *Resource types and description*

Type	Description
GET	Gets the value attribute of the resource
SET	Sets a given new value as the data value of the resource
ACT	Invokes the resource (method execution) with a set of defined parameters
SUBSCRIBE	Subscribes to the resource with optional condition and periodicity
UNSUBSCRIBE	Removes an existing subscription

Table 6.3. *Resource access methods*

The access methods that can be associated with a resource depend on the resource type. For example, a GET method can only be associated with resources of Property, StateVariable and SensorData types. A SET method can only be associated with StateVariable and modifiable Property resources. An ACT method can only be associated with an Action resource. SUBSCRIBE and UNSUBSCRIBE methods can be associated with any resources.

6.4. Platform security management

The sensiNact platform is based on the OSGi Framework. A first level of security is reached by way of some available security "tools" in this environment: *ServicePermission* and *ConditionalPermissionAdmin*. The *ServicePermission* is a module's authority to register or use a service;

– the *register* action allows a module to register a service on the specified names;

– the *get* action allows a module to detect a service and use it.

Permission to use a service is required to detect events regarding the service. Untrusted modules should not be able to detect the presence of certain services unless they have the appropriate *ServicePermission* to use the specific one. The *ConditionalPermissionAdmin* is a framework service used to administer conditional permissions that can be added to, retrieved from and removed from the framework.

The sensiNact platform defines service permissions in such a way that access to the ones it provides is forbidden except if a specific condition is encountered (a sensiNact-specific conditional permission), this condition being that the client is the sensiNact *SecuredAccess* service. sensiNact services also have to use *SecuredAccess* to be able to "talk" to each other; modalities of such exchanges depend on the *UserProfile* of the user of these services (the user can be the system itself). A *UserProfile* can be defined at each level of the hierarchical sensiNact resource model: *ServiceProvider*, *Service* and *Resource*. Five *UserProfiles* have predefined access rights: Owner, Administrator, Authenticated, Anonymous and Unauthorized.

When asking for a data structure of the sensiNact resource model, the access rights of the user are retrieved (see Figure 6.6); the set of this user's accessible AccessMethods for the specific data structure is built and returned as part of the description object. Each future potential interaction of the user on the data structure will be made by way of this description object. For a remote access, a security token is also generated and transmitted to the user, to avoid repeating the security policy processing. A token is defined for a user and for a data structure (and so for a previously created description object).

The Security & Dependability functional block is used for authentication and to retrieve identity material from which it will be possible to associate a user and a sensiNact resource model data structure to a *UserProfile*.

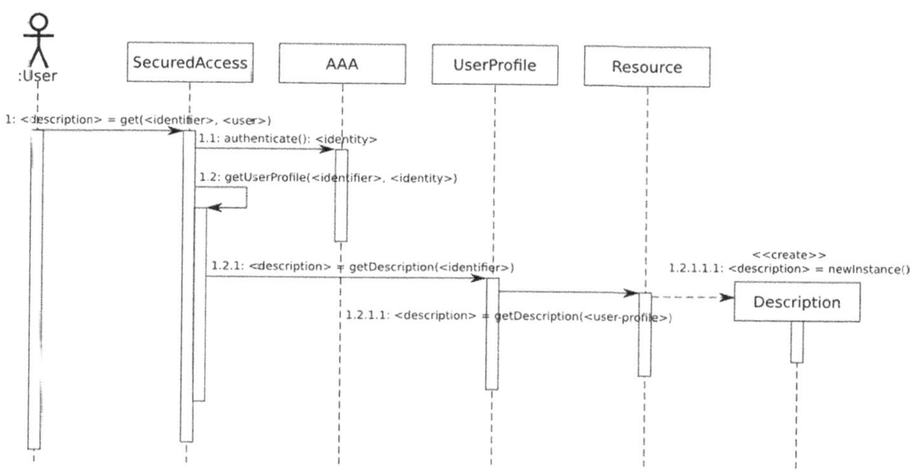

Figure 6.6. *Secured access sequence diagram*

In addition to the database managed by the Security & Dependability functional block, used to authenticate a user and to retrieve its identity in the system, the sensiNact platform manages an internal database allowing users to link this identity to a *UserProfile* for a specific data structure. For all data structures for which the user has not been registered, the *Anonymous UserProfile* is used by default (except if the owner of a resource has defined this default profile to another one). The internal database also gathers information relative to the minimum required *UserProfile* to access to data structures. This definition can be made at each level of the resource model, knowing that if no *UserProfile* is defined for a data structure, the one specified for its parent is used.

For example, according to Figure 6.7, a user trying to access the *Service ProviderX* for which its *UserProfile* is *Anonymous* will receive a description object in which only one *Service* will be referenced (*ServiceX1*), containing a single *Resource* (*ResourceX1S2*) providing two *AccessMethod*s, GET and SUBSCRIBE.

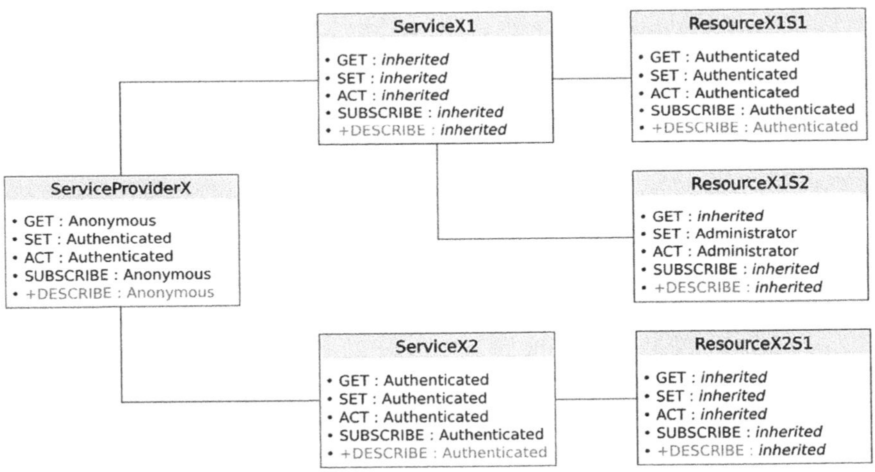

Figure 6.7. *Access right inheritance diagram example. This example shows that access rights are inherited from the parent node, i.e. the service provider, unless they are specifically specified*

6.5. The sensiNact studio

The sensiNact Studio is a key element to interact with the devices and to ease the creation of applications. The Studio is based on Eclipse platform [DAN 05] and built as a rich-client platform application. The Graphical User Interface (GUI) is developed using the views mechanism from Eclipse. Thus, it proposes many view related to browse devices, locate devices on a map and interact with devices, i.e. obtain value from sensors or perform an action on an actuator.

The Studio is also dedicated to the creation of IoT application without particular knowledge of any specific software development language. Written in a Domain Specific Language (DSL), an IoT application includes an Event-Condition-Action (ECA) rule [DEN 89], which defines the behavior of devices in the case of the occurrence of event and the verification of one or several conditions.

The following sections detail the GUI and the tool to create sensiNact applications.

6.5.1. *Graphical user interface*

The GUI includes different views: navigator, deployment, properties views, as well as a Domain-Specific Language (DSL) editor and a Graphical Description Language (GDL). Figures 6.8 and 6.9 provide a screenshot of the GUI.

The navigator view (see Figure 6.8) shows active devices in the environment. This view is automatically updated when devices join or leave the environment. Moreover, in the navigator, the user can find services provided by a device and for any service. He can also list the resources exposed by the latter. Resources represent information (properties, state variables, sensed data) or functionalities (actions) exposed by services. For example, if we consider temperature service, it provides a resource of temperature offering temperature measurements, on demand, periodically or if a certain condition is encountered. The obtained measurements can be graphically visualized. Similarly, a light service example has three action type resources: *TURN_ON*, *TURN_OFF* and *DIM*. These actions can be performed remotely. More generally, based on the sensiNact platform service/resource model, the tool provides performing the generic sensiNact methods remotely on the exposed resources: GET, SET, ACT, SUBSCRIBE and UNSUBSCRIBE.

The deployment view shows the location of indoor and outdoor devices. This helps the user to visualize all the available IoT devices in the system and to use them for the development of their applications. The locations of the devices can be updated by the user. The property view gives more information about manipulated entities in the GUI and allows the user to modify them. The GUI includes two other views, namely the DSL editor and GDL editor (see Figure 6.9), which will be detailed in the following sections.

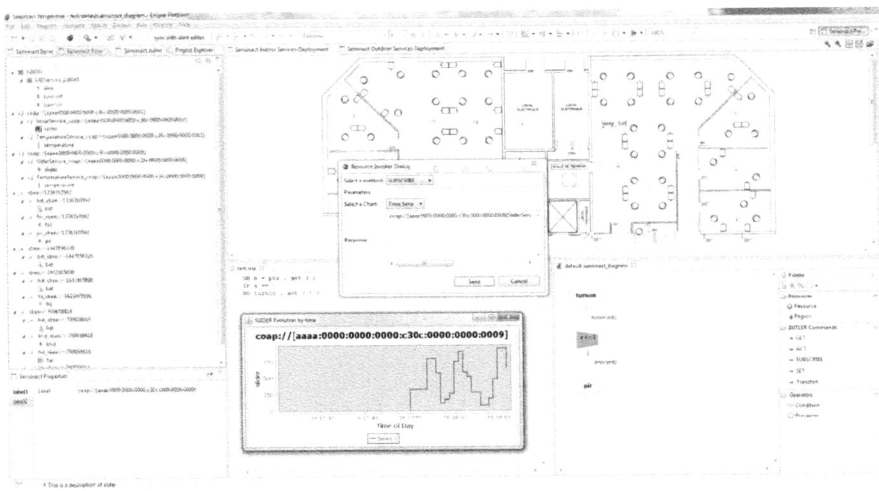

Figure 6.8. *The navigator view (left), the deployment view (top right) and the property view (bottom left). For a color version of the figure, see www.iste.co.uk/elfallah/smartcities.zip*

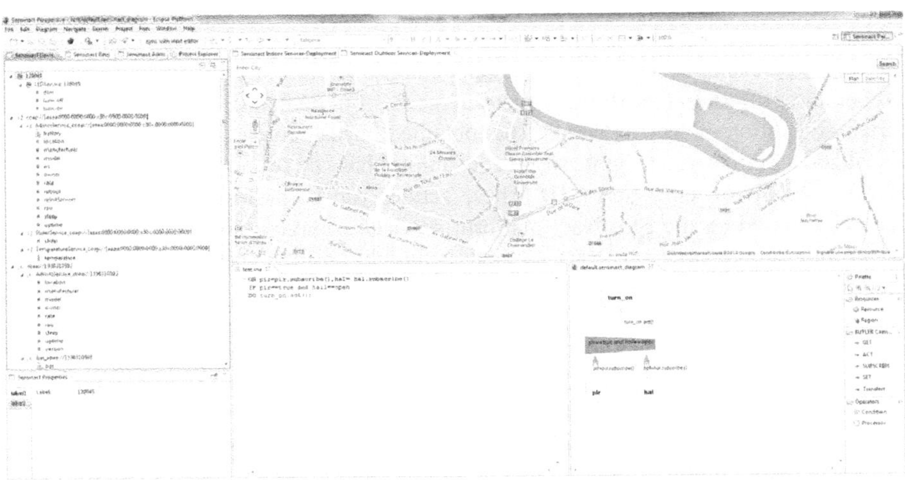

Figure 6.9. *DSL editor (middle bottom) and GDL (bottom right). For a color version of the figure, see www.iste.co.uk/elfallah/smartcities.zip*

6.5.2. *Creating applications*

The Studio eases the development of applications for anyone. It enables us to create applications based on the resources provided by the sensiNact data model using a composition of them. The Studio provides a way to develop event-driven applications i.e. based on the ECA axiom. This axiom assumes that the application is triggered when the required events occur. Then, if all conditions are satisfied, the actions are performed. Generally, in an IoT environment, events are created from sensors, and the actions are performed using the available actuators.

6.5.2.1. *The sensiNact DSL syntax (sNa syntax)*

The Studio comes with a Domain-Specific Language (DSL) based on the ECA axiom, i.e. the sNa syntax, to deal with the event-driven IoT environment. The developer uses the information available at runtime to compose its application. The sNa syntax is composed of the following block (and summarized with an example in Table 6.4):

[resource <resource>]+	resource shortcutLightOn=/URI/of/the/lightOn resource shortcutLightOff=/URI/of/the/lightOff resource shortcutButton=/URI/of/the/button
on <events>	on shortcutButton.subscribe()
[if <condition> do]+	if shortcut.get() == true do
[<actions>]+	shortcutLightOn.act()
[else do]?	else do
[<actions>]?+	shortcutLightOff.act()
end if;	end if;

Table 6.4. *sNa syntax using ECA and a simple example applied to a button and a light. This example turns on a light when the button state is "true" and turns off the light otherwise*

– *The shortcut block*: each resource is accessible through a unique URI. This block aims at creating shortcut for the next blocks. The keyword to create a shortcut is *resource* followed by the shortcut name and the URI of the resource.

– *The event block*: the developer defines which resources the application is processed on. When the event is thrown and corresponds to the expected event, the application goes to the next block. The keyword to create this one-line block is *on* followed by the shortcut of the resource and the keyword *subscribe*. Subscription to a resource can also be conditional.

– *The conditional block*: once the application has been triggered, and before the action can be done, the data from the resource have to satisfy the conditional block. The keyword for this block is *if* followed by the condition to validate.

– *The action block*: if the conditional block is satisfied, actions are performed in the order that they are listed. The actions can be physical actions on actuators or virtual actions such as changing the format of a data using a mathematical function. The available actions, also named functions, are listed below:

- *Basic functions*: addition, subtraction, division, multiplication, modulo, string concatenation, substring, various operators (e.g. equals, greater than, lesser than, different), ACT and SET methods on sensiNact resources.

- *Complex Event Processing functions* using the Esper library [ESP 16]: after, before, coincides, average, average deviation, count, max, min, median, standard deviation and sum.

6.5.2.2. *Graphical description language*

GDL (Graphical Description Language) is a graphical description corresponding to the ECA language described in the previous section. The graphical representation follows the same logic of device description to facilitate application development with sensiNact environment. The user can find resources, access methods and condition entities in the tool bar of the GDL (right-hand side of Figure 6.9) to link the resources between them with conditions. The DSL and the GDL editors are synchronized automatically to help the developer to choose both ways of creating the applications.

To develop an application with the GDL, the user can use *Drag and Drop* functionalities from the deployment view, thus allowing the user deciding which devices to use to develop its application. The user can also use the GDL tool bar to add a device resource, condition and connections between them. The property view can be used to assist the user defining condition details (this functionality uses Xtext [MOR 10] capability in the property view).

6.5.3. *Application deployment*

The sNa syntax is a human readable syntax, showing concepts and relationships between them. However, the platform needs an easy to parse syntax, which should be less verbose to be easily transmitted to the platform over a network. For this reason, we have followed the approach summarized in Figure 6.10.

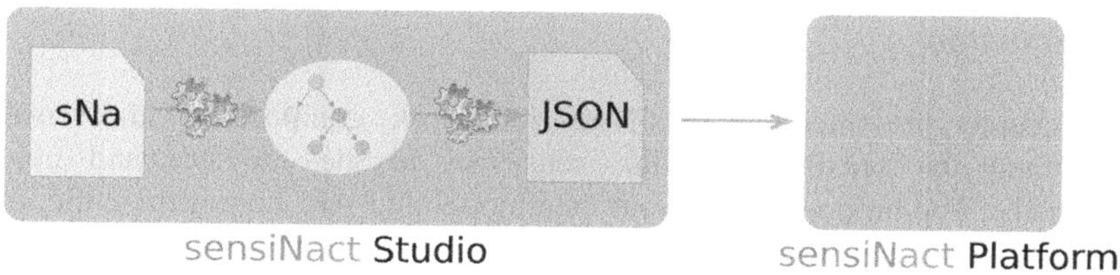

Figure 6.10. *Application deployment. An application is written using the sNa syntax, then it is converted into a JSON file and sent to the platform that uses and runs it. For a color version of the figure, see www.iste.co.uk/elfallah/smartcities.zip*

On the left of the figure, the sNa file is the one directly edited in the studio. Thanks to Xtext, this file is transformed into graph of Java objects. This graph is then processed to generate a JSON file, finally transmitted to the AppManager, i.e. the service in the platform in charge of managing the applications, hosted by the platform.

The AppManager provides a lifecycle to manage the applications. It enables the user to process various checks during different steps of the lifecycle of the application (e.g. application description consistency, resources permissions). The first step is to install the application, i.e. send the application description. If there is a problem, the AppManager returns an error. Once the application is installed, it can be started and its state changes to "Resolving". If there is a problem during this step, the application enters in the "Unresolved" state. Otherwise, the application is active until it is stopped or an exception occurs (see Figure 6.11).

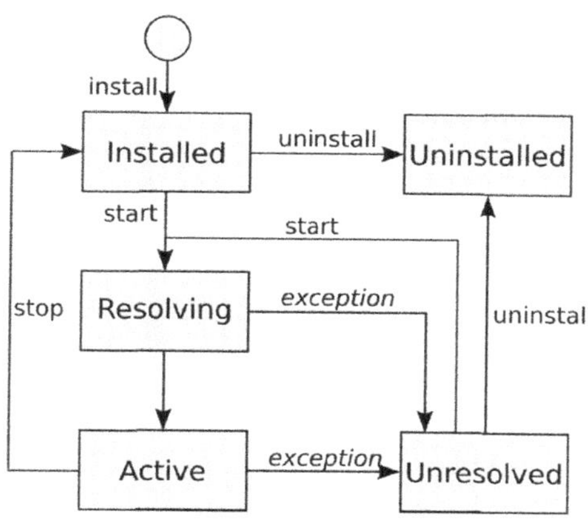

Figure 6.11. *Lifecycle of an application*

6.6. Conclusion

This chapter presented the sensiNact framework, composed of the sensiNact platform and the sensiNact Studio. sensiNact handles devices and network heterogeneity. The proposed sensiNact data model and API homogenize the access to those environments, unlocking the transversal applications to be developed in domains such as smart home/building, smart health care, smart city, smart transportation etc. The easy-to-use, generic and extensible data model eases the development of applications by third-party users. The right access mechanism of sensiNact provides a full control on the data transmitted by the sensors and on the actuators available in the environment. Moreover, sensiNact Studio provides an IoT application development, deployment and management tool, which allows developers rapidly prototyping IoT applications, without requiring any particular knowledge about the underlying IoT technology.

sensiNact has been successfully used in close to real-life environments in various application domains such as smart city, smart home, smart shopping, smart health care within the context of collaborative projects such as BUTLER[1], ClouT[2] (see Chapter 5), FESTIVAL[3], SocIoTal[4] and OrganiCity[5]. Within the context of those projects, in particular in the smart city domain, sensiNact has been already used to access data from various European and Japanese cities, such as Santander, Genova, Fujisawa and Mitaka. It is currently expanding the number of supported cities thanks to ongoing and future collaborative and bilateral projects with additional cities worldwide.

6.7. Bibliography

[ALL 12] Alliance OSGi., OSGi Service Platform, Core Specification, Version 5, 2012.

[DAN 05] D'ANJOU J., *The Java Developer's Guide to Eclipse*, Addison-Wesley Professional, 2005.

[DEN 89] DENNIS MCCARTHY U.D., "The architecture of an active database management system", *ACM Sigmod Record*, pp. 215–224, 1989.

[ESP 16] ESPERTECH, http://espertech.com/, 2016.

[EXO 16] EXOSITE, https://exosite.com/, 2016.

[HOB 16] HOBSON, http://hobson-automation.com, 2016.

1 www.iot-butler.eu.
2 http://clout-project.eu.
3 www.festival-project.eu.
4 http://sociotal.eu.
5 http://organicity.eu.

[IBM 16] IBM Bluemix, http://www.ibm.com/cloud-computing/bluemix/, 2016.

[IOT 16] IoTSens, http://www.iotsens.com, 2016.

[KAA 16] Kaa, http://www.kaaproject.org/, 2016.

[KUR 16] Kura, https://www.eclipse.org/kura/, 2016.

[MAC 06] MacKenzie C.M., Reference model for service oriented architecture 1.0, OASIS standard, 2006.

[MOR 10] Moritz Eysholdt H.B., "Xtext: implement your language faster than the quick and dirty way", *Proceedings of the ACM International Conference Companion on Object Oriented Programming Systems Languages and Applications Companion*, pp. 307–309, 2010.

[OPE 16a] Open the Box, http://openthebox.org, 2016.

[OPE 16b] OpenHAB, http://www.openhab.org/, 2016.

[THI 16a] Thinger, https://thinger.io/, 2016.

[THI 16b] ThingWorx, http://www.thingworx.com/, 2016.

7

Verification and Configuration of Smart Space Applications

7.1. Introduction

Smartness has been the key concept in the continuous advancement of embedded devices that autonomously provide increasingly powerful sensing and actuating capabilities. Systems detect the context of users through sensing capabilities. Then, with the power of flexible and intelligent software control, they affect users or their surrounding environment with actuating capabilities. Such smart home applications have been investigated within different paradigms, originally within ubiquitous computing and pervasive computing [WEI 99, SAT 01]. Smart space applications have recently been attracting wider attention with the enhancement of device capabilities and increase in demand for smartness (Internet of Things and smart cities).

Smart space applications have a direct impact on human and social activities, which not only enhance attractive features but also increase the complexity of systems as well as demand for dependability. Although major research activities often start with facilitating the running of applications with frameworks and middleware, it is also essential to tackle difficulties to ensure dependability in intricate software for smart space applications.

One of the specific difficulties in smart space applications is the event-driven nature of device control. An application system senses changes in contexts and actuates proper reactions to the changes to support user activities or prevent undesirable situations. As there are enormous possibilities of event sequences, it is easy to overlook a certain situation and fail to configure the activation of proper

Chapter written by Fuyuki ISHIKAWA and Shinichi HONIDEN.

actions or the prevention of improper actions. This difficulty is increased by the existence of conflicts between multiple applications for multiple users with multiple devices.

In this chapter, we will discuss issues in tackling this challenge in smart space applications. This is not only related to verification of design by engineers but also to configuration by space administrators or end users. Traditionally, model checking techniques have been considered effective for problems regarding various possibilities of state transitions, which include potential conflicts that come to the surface only in specific situations [BER 01]. However, it is necessary to properly set up model checkers by properly defining state spaces to be explored and properties to be verified for effective verification.

We propose a framework for verifying and configuring smart space applications. The specific questions addressed in this chapter are as follows:

– *how can we model the physical effects of devices on users?* This is regarding the semantic representation of smart spaces, e.g. "a user hears a sound from a speaker";

– *how can we extract the essences of state changes, avoiding unnecessary complexity with implementation details?* This is regarding the construction of the state space to be explored using model checkers;

– *what are the requirement patterns of smart spaces that consist of controllable applications and uncontrollable users?* This is regarding the definition of properties in the form of mathematical formulas to be verified using model checkers;

– *how can we connect these technical foundations to development environments for smart spaces?* This is regarding the implementation of a practical tool.

We also discuss a case study to illustrate how potential conflicts are detected and resolved in smart spaces. We believe our approach in the framework provides the essential foundation for the dependability of smart space applications.

The remainder of this chapter is organized as follows. We discuss the conflicts in smart space applications in section 7.2. We describe the proposed framework for verifying and configuring smart space applications in section 7.3. We discuss the use of the framework with case studies in sections 7.4 and 7.5 and give concluding remarks in section 7.6.

7.2. Conflicts in smart space applications

7.2.1. *Event-driven control of smart spaces*

Context awareness has been considered one of the primary characteristics for smart space applications. Context represents "any information that can be used to characterize the situation of an entity" [DEY 01] and is, thus, essential for application systems to be smart for the target entities, typically human users. Context-awareness leads to various features, such as timely recommendation of activities according to the user context.

In this chapter, we will focus on event-driven control actions of smart spaces as one of the most essential features. An application system senses changes in the context and properly reacts to support user activities or prevent undesirable situations. This type of application not only uses sensor devices but also actuator devices that directly affect the real world.

7.2.2. *Description of event-driven behavior*

One common way to define event-driven behavior is using event condition action (ECA) rules [DIT 95]. An ECA rule has the form of "ON *event* IF *condition* DO *action*", which defines an *action* triggered to react to an *event* only under a certain *condition*. For example, a rule for an air-conditioning application activates the cooling function in response to detection of a temperature increase beyond the threshold, only if there are people in the room.

The ECA rules have been widely used for describing the control behavior in studies on smart spaces [RAN 03, HEN 06, TWI 09]. Recent studies have investigated sophisticated integrated development environments (IDEs), not only the textual language aspect of ECA rules. The sensiNact environment is one such IDE and targets event-driven control of Internet-of-Things (IoT) services [CAN 14]. It leverages graphical user-interface features such as code completion or diagram-based input of ECA rules. It also provides easy integration of IoT services into the rules through service management and visualization capabilities.

7.2.3. *Conflicts in event-driven control*

Validation of ECA rules is essentially difficult because there can be various execution sequences depending on what events occur and the order of their occurence. Specifically, it is easy to overlook conflicts between rules configured for different application requirements.

As a simple scenario, consider a smart office system that supports the presentation of slides and electronic posters, the demonstration of tools and discussion in a room (Figure 7.1). This system is expected to support both presenters and audiences, often without explicit commands from them, while preventing undesirable situations. In this section, a very small part is discussed to quickly illustrate the difficulties with ECA rules.

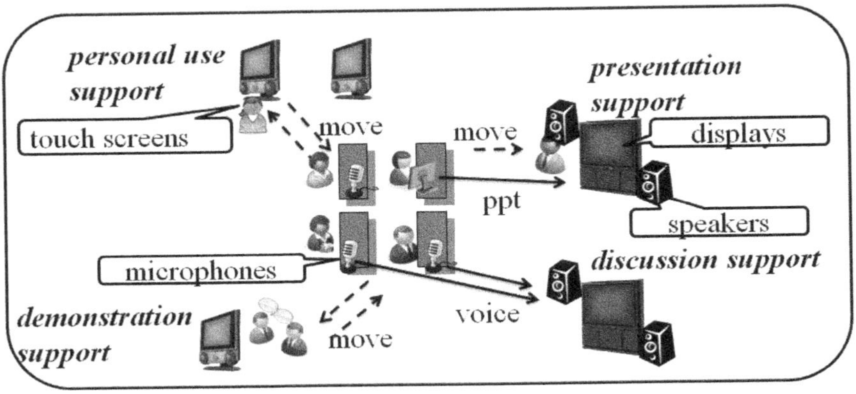

Figure 7.1. *Smart office system*

An example of specifications of this system is shown in Figure 7.2 regarding simple usage of shared displays. It includes requirements on system R1 and R2, as well as behavior specification (ECA rules) to meet the requirements, B1 and B2.

R1 An authorized user can use a nearby display to browse shared business information such as a shared calendar.

B1 Given a request by an authorized user nearby (who is allowed to access the information and *is able to see the display*), the system starts to show the requested information on the display.

R2 Shared business information is never seen by an unauthorized user.

B2 The system stops showing the information when an unauthorized user is nearby (*and can see the information on the display*).

Figure 7.2. *Example specifications with deficiency*

The example specifications are not satisfactory in the sense that the set of behavior specifications B1 and B2 do not meet requirement R2. In fact, behavior B1 can start to show the information on the display even when there is already an unauthorized user there. This situation means that there is a conflict between R1 and R2, i.e. they cannot be met as they are (without any restrictions). If a decision is made to put higher priority on R2, B1 and R1 are then modified by adding a constraint: "*only if there is no unauthorized user nearby*".

The problem here is how we can detect this conflict. Suppose we find this conflict through test scenarios, either executed in the physical environment or in a simulation model (or even in the engineer's mind). This means that we need to include a specific test scenario, among many possible scenarios, in which there is an unauthorized user nearby and an authorized user making a request.

To summarize, the above scenario demonstrates a point that there are often conflicts between requirements on different aspects, e.g. availability versus security. Such conflicts come to the surface only in specific situations or event sequences; thus they are likely to be overlooked by engineers, and it is essential to have automated, systematic support to detect such situations.

The above scenario also demonstrates a case with multiple requirements or applications. In more general scenarios, conflicts can also occur due to the existence of multiple users and multiple devices. For example, a smart audio-visual application may start playing a video on a TV in the living room when a user moves into the room while playing the video on his mobile phone. However, this causes a conflict if another user has been using a speaker in the same living room for a phone call (without a proper guard condition to avoid this situation).

This nature of multiple applications, users and devices is essential in smart spaces. For example, the study in [MUN 14] demonstrated 35 example apps for a smart home. They include control of devices such as lights, speakers and windows in response to a change in time, humidity, energy usage and so on.

Note that the correction to avoid conflicts can be basically considered as verification and debugging activities by professional engineers at the design phase. In more advanced scenarios, it is possible to imagine space administrators or end users installing and configuring the applications. They should be properly configured by being aware of conflicts according to their policies and preferences. It is easy for them to carry out this configuration in the form of priorities over applications or rules. In this chapter, we use ECA rule description based on conditions and do not give numeric values of priorities, as they can be converted in a somewhat trivial way.

7.2.4. *Application of model checking techniques*

Model checking is a well-known, effective approach for problems with various possibilities of state transitions and conflicts that come to the surface only in specific situations [BER 01]. Model checking explores possible state transitions to ensure that the given properties hold or detect counterexamples in which the properties do not hold. Tools with efficient algorithms as well as practical configurability have been implemented such as SPIN [SPI 15]. However, there are issues to address for effective use of model checking techniques, as discussed below.

First, it is necessary to construct state spaces to be explored. Generally, this is done with a language provided by a model checker, e.g. the PROMERA language for the SPIN tool. However, it is not cost-effective to require the involved individuals to learn and use such a language, especially requiring encoding of the problem domain structure into the structure of the language (e.g. ECA rules into state transitions).

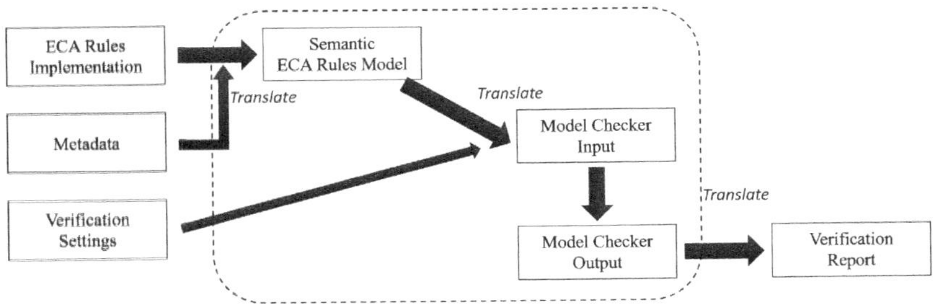

Figure 7.3. *Overview of proposed framework*

This encoding is not a simple labor-intensive task but rather a key for effective and efficient verification. It is necessary to extract and model the essences of the application behavior as well as possibilities of changes in the environment, including users. This is the requisite for meaningful verification with acceptable performance, avoiding exploration of enormous possibilities that do not essentially affect the conflicts.

Last but not least, support for specification of properties is also preferable. Formula expressions in temporal logic are used in model checking, which can be supported by typical property patterns [SPE 15] (e.g. a property pattern "*A* can occur only before *B*"). Besides such generic support, domain-specific in-depth support will be attractive for smart space applications. Specifically, with the existence of uncontrollable users, who may do good or bad, it is necessary to clarify assumptions or expectations on users to discuss what the system should do.

7.3. Framework for verifying and configuring smart space applications

7.3.1. *Overview*

Figure 7.3 illustrates the proposed framework for verifying and configuring smart space applications. The left side shows three elements of the input and the right side shows one element of the output. The dashed rectangle denotes the boundary of the tool: users of the framework do not need to look inside it. This architecture is defined to bridge the gaps between practical domain-specific representations for smart space applications and those for model checkers. We start with the ECA-based internal representations i.e. the *Semantic ECA Rules Model* element in Figure 7.3.

7.3.2. *Semantic model*

An abstract model for ECA rules in smart spaces is defined to represent the semantics of application behavior (the *Semantic ECA Rules Model* element in Figure 7.3). This model is focused on the essential states of the smart space, abstracting away implementation details such as communications between devices and controllers. This model can be considered as a design model directly specified by engineers or extracted from a program in the form of ECA rules and related metadata (described later in section 7.3.3). For specification of the semantic model, a domain-specific modeling language is provided in the framework.

7.3.2.1. *Scope-based modeling*

States of smart spaces are modeled with the concept of *scopes*. A scope denotes a boundary to distinguish if a device can affect a user [SAT 07]. A user can conduct a certain type of physical interaction (e.g. see) with a target device only if he/she is inside the scope and the device is on for the interaction. A scope exists for each pair of a user and certain type of physical interaction (e.g. visual) of a device.

This scope corresponds to the term "nearby" in the natural language example in section 7.2.3. This representation allows for analysis without implementation details. For example, it abstracts away details on numeric distances or configurations of sensor capabilities. It also abstracts away how the scope changes, e.g. by increasing the volume of a speaker or by letting the user enter a certain area. It should be noted that this modeling approach takes into account virtual scopes that can flexibly correspond to various concepts in the real world. It can be used to follow a simple modeling approach to consider only physical areas such as rooms or buildings.

Figure 7.4 illustrates a few examples of scope-based modeling, as described below:

– the left figure denotes the situation of the example scenario, where a user comes close and is able to see the display that has been activated for another user. This is explained by inclusion of the two users in the scope for visual interaction with the display device;

– the one in between denotes a situation of sound conflicts, where a user hears different sounds from different audio devices and becomes uncomfortable, e.g. when a movie player is activated automatically while a recipe reader is running in a smart home application. This is explained by inclusion of the user in the overlapping two scopes for audio interaction with the two different devices;

– the right figure denotes a situation in which a user sees different direction instructions in a smart museum application. This is explained similarly by inclusion of the user in the two overlapping scopes for two visual devices.

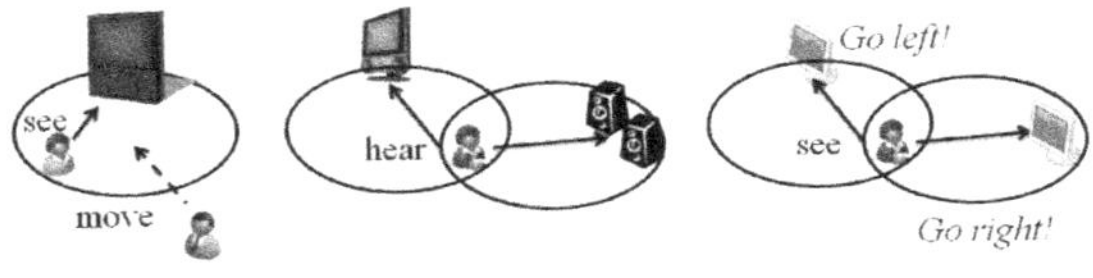

Figure 7.4. *Modeling of scopes for physical interactions*

These examples include conflicts that can occur depending on the relationships between users and scopes (i.e. user inclusion), or between scopes (i.e. scope overlap). By modeling and examining such relationships explicitly, implicit assumptions on device layout or potential conflicts can be clarified.

7.3.2.2. *Abstract ECA rules*

An example of abstract ECA rules in the semantic model is given below: "when a user enters $room1$ and if device $speaker1$ is off, then turn it on", which results in some kind of sound.

```
eca1{
  ON getIn(USER, room1)
  IF isOff(speaker1)
  DO on(speaker1)
}
```

In this rule, *getIn*, *isOff* and *on* are vocabularies provided by the framework. The *USER* keyword denotes any human user, not a specific person. Another example of vocabularies in the framework is specification of events or conditions on the number of users in a scope. Thus, events, conditions and actions are described in terms of scopes and device states for an abstract model without implementation details.

The above example rule shows the simplest form of the rule specification language in the framework. The language actually provides foundational syntax elements, including the following:

– operators regarding events, conditions and actions. For example, logical operators to combine multiple events or multiple conditions and specification of atomic actions;

– definitions and references to application-specific state variables. For example, a variable that represents user roles such as "administrator", and a variable that represents types of content used by devices such as "confidential";

– arguments of standard functions. For example, the vocabulary for actions on can specify what types of physical interactions are activated and what data are used, e.g. "activate the device for the audio interaction to play the content type "music".

It should be noted that it is not always necessary to model details, even if they are expected to exist in a real system. For example, which music files to play may not matter when we want to discuss whether sound conflicts can occur.

7.3.2.3. *Effect definitions*

In fact, the rule in the above example does not state that `speaker1` makes sound. It is necessary to define each device or its service in terms of what effects it produces on the environment or user. With this definition, the abstract ECA rules denote the semantics of the application behavior. The following rule defines one device that enables the *audio*-type interaction with users, i.e. users can hear sounds from the device when they are in the scope of the device.

```
Devices = { speaker1 : [audio] }
```

7.3.2.4. *Environment definitions*

Assumptions on the relationships between scopes can be defined. In other words, by default, all scopes can overlap at runtime, leading to situations such as sound conflicts (illustrated in Figure 7.4). Thus, environment definitions can be given to explicitly state the disjointness of scopes. In simple cases, this can be done by defining hierarchical structures of physical spaces such as buildings, floors and rooms, then defining scopes in terms of spaces. For example, suppose sounds from a device are effective in one room and sounds of another device are effective in the adjacent room. The scope of the two devices for the audio interaction is then defined as disjoint.

7.3.2.5. *User behavior*

Actions by users affect the environment, not only actions from the system. It is necessary to distinguish them because users are not controllable when we try to define proper control of the system.

In the example below, when a user leaves the scope of the device, the system asks the user to move back into the scope. If the user accepts, the system state just continues as it was. If the user denies, the system shuts the device down. In the action specification, the symbol `::` represents alternative actions possibly with guard conditions, and in this case the keywords `accept` and `deny` are used to specify conditions in terms of user agreement (with an identifier `req1`).

```
eca1{
  ON getOutScope(u, dev0, visual)
  IF isOn(dev0)
  DO
      :: req1/accept: getInScope(u, dev0, visual)
      :: req1/deny: off(dev0)
}
```

From the viewpoint of state transitions, this description just means two possibilities that the application cannot control; thus, the model checker should try both. This kind of description is referred to when properties are discussed under assumptions on user behavior, as described in the following section.

7.3.3. *Definition of state transition model*

The semantic model presented in section 7.3.2 denotes the application behavior, which also defines the states explored by the application. The states consist of device status (*on, off*), relationships between each user and each device (*in, out*) and application-specific variables. The transitions between these states are defined by the abstract ECA rules. Interpretation of the transitions defined by the rules is done using the "model checker input" element in Figure 7.3.

7.3.3.1. *Abstraction*

Even though the behavior defined by the ECA rule is intuitively clear, there are multiple possibilities to map it with state transitions. Specifically, there are choices about what is considered as one step and how different rules are interleaved.

One extreme example is to construct state transitions so that all event occurrences, condition evaluations and action activations can take some time and, thus, are interleaved by each other. Suppose a transition occurred that denotes an event "user1 entered room1". The possibilities of the next transitions not only include the activation of a responding action by an ECA rule but also interleaving occurrences of other events. It is even possible to have a scenario in which "the user moves in and soon out, while the system's reaction to the first inward movement is activated when the user is already out" (thus, say, the light is turned on even though there is no one there). This model allows for analysis of failures in very pessimistic scenarios. However, this is probably too granular or implementation-specific for early analysis and too pessimistic with scenarios that do not actually occur with assumptions on negligible execution time of ECA rules.

The proposed framework by default uses the atomic execution of each ECA rule, in a sense it is not interleaved by other rules or other events. In other words, the state transitions are constructed to repeatedly simulate a cycle of an event occurrence and evaluation of responding ECA rules. This model allows for analysis of conflicts between the intentions of multiple rules, without taking into account implementation-level issues.

On the other hand, the above semantics does not mean the atomicity of an event occurrence and execution of the responding ECA rules. For example, suppose a rule to turn off the display when an unauthorized user comes nearby. There is still an intermediate state in which the display is still on but the unauthorized user is nearby. It is necessary to consider such a state to be aware of the fact that ECA rules responding to undesirable situations are *a posteriori* means. This is distinguished from an alternative means of prevention to lock the door before the display with confidential content is turned on.

7.3.3.2. *User behavior*

The state spaces we have been discussing are for exhaustive exploration by model checkers. For this purpose, users are modeled as “monkeys” that can take any possible action in any situations: thus, testing the application exhaustively. Specifically, a user can move freely but under constraints defined by scope relationships in the environmental setting. For example, a user in a certain scope can enter another scope (without leaving the original scope), only if the two scopes are said to “always overlap” or “sometimes overlap”. Events of user requests are also generated to cover all possibilities of user actions (`accept` and `deny`).

The number of users in the modeled space matters in terms of what possibilities are to be explored. In the extreme case, if we consider only one user in the space, we cannot detect the conflicts caused only by multiple users, such as the conflict in the example. On the other hand, it is not practically feasible to include as many users as actual, e.g. 1,000 unique users, as it leads to meaningless state explosion. It is necessary to estimate a sufficient number of users that can cause different kinds of conflicts. For example, it can be enough to check only four users to be confident when conflicts are imagined with multiple (two or more) users.

It is possible to configure the number of users when running model checking, in the *Verification Settings* element in Figure 7.3.

7.3.4. *Properties to verify*

The *Verification Settings* element includes the definition of what properties are verified by model checking. Temporal logic formulas can be given directly, as in general model checkers. We only discuss linear temporal logic (LTL), which is used in the SPIN tool [SPI 15] used in the current implementation.

In LTL, temporal operators can be used such as “in future or eventually (**F**)” and “globally or always (**G**)”, besides general operators in propositional logic. Model checkers verify whether a property holds for all possible execution paths. Natural language expressions or easy interfaces (e.g. pull-down menu) may be used with typical patterns [SPE 15]. By not only relying on the general patterns, the framework provides support for domain-specific properties to be checked in smart spaces.

7.3.4.1. *Typical properties*

For example, the framework allows for easy specification of the property pattern of which sounds from two different speakers are never received by a user at the same time (here “received by a user” means “a user is inside the scope and the device is on”).

The framework can generate an LTL formula given the specification of the target audio devices.

7.3.4.2. *Properties with human users*

The proposed framework also includes two specific patterns of LTL formulas by taking into account the existence of human users in smart spaces.

Generally, safety properties require that undesirable situations are never reached. A safe property is written as $\mathbf{G}\neg P$, where property P denotes the undesirable situations. However, they are often nonsense in smart spaces, as the application behavior cannot prevent undesirable situations caused by uncontrollable user actions.

In the framework, traditional safety properties are renamed *strong prevention* to clarify the meaning. Another property pattern *weak prevention* is introduced to mean "an undesirable situation will always eventually be stopped", namely, $\mathbf{G}(P \Rightarrow \mathbf{F}\neg P)$.

However, this property would still be unsatisfactory, if P is stopped just because a user eventually takes some action while the system does not have any reaction to stop P. We can avoid this situation and verify the system behavior by including the possibility that a user remains doing nothing infinitely. In that case, the weak prevention property does not hold if the application behavior lacks corresponding reactions to stop P.

Another point is that there are properties that hold only if users agree with a certain type of request from the application. This is often the case in which the system provides some attractive functionality to proactively and automatically support the users, but it cannot force the users to accept the functionality. In such a case, verified properties should state correctness of the functionality under the assumption that the users do not deny it. Such an additional pattern of user agreement, for any P of prevention, fairness and others can be defined as $(\mathbf{G}\neg A) \Rightarrow P$, where A means occurrence of user denial identified by the labels in the specification.

These typical properties can be easily specified in the framework by only indicating parameters such as the definition of undesirable situations (P in the above description).

7.3.5. *Implementation*

7.3.5.1. *Translation from practical ECA rules implementation*

The semantic model is a direct representation of the semantics. It can be used as a high-level language for specification or design in early phases of the development process. Another use is to verify a program in a programing language for ECA rules.

The element *ECA Rules Implementation* denotes this practical language, e.g. sensiNact mentioned in section 7.2.2.

As the objective of such languages is detailed design or implementation, it often uses references to components or services in description of events, conditions and actions. For example, the action to start a display can be written as service invocation, e.g. `speakerService1.play.act()`, linked with service description (e.g. REST service access point). This description does not formally define any relationships between a specific speaker device or its state (is it making sound or not). Thus, it is necessary to map to the semantic model by attaching metadata on each device or service.

We exemplified this concept with the sensiNact platform. The foundational part (core behavior of ECA rules) can be automatically translated by defining the metadata in the approach mentioned above. When advanced functionality is required (e.g. distinguishing user actions), it is necessary to annotate the original rules with an extended syntax imported from the language for abstract ECA rules.

7.3.6. *Model checker implementation*

The current implementation uses SPIN [SPI 15], one of the most popular model checkers. In many cases, it is possible to hide the SPIN tool from the users of the framework, as shown in Figure 7.3. For this purpose, backward translation is also provided on the verification result (*Verification Report* in Figure 7.3). Manual operation on SPIN is required when it is necessary to give detailed configuration on the tool or to use its various functions (e.g. interactive simulation).

7.4. Case study

We describe a case study that illustrates the use of the framework for verifying and debugging ECA rules. The example mentioned in section 7.2.3 is revisited in more detail.

7.4.1. *Scenario and initial specification*

Figure 7.1 illustrates part of the system described for supporting event operations, such as an academic conference or an open house. The space can be used in two modes for presentation and demonstration. In the presentation mode, a presentation is made for all the attendees in the space, using the largest display and an embedded speaker system to cover the space. In the demonstration mode, demonstrations are supported to be presented using small displays and demonstrations around the space, with often automatic activation of each demonstration when an attendee moves nearby. For simplicity, only one set of display and speaker is considered for the demonstration mode.

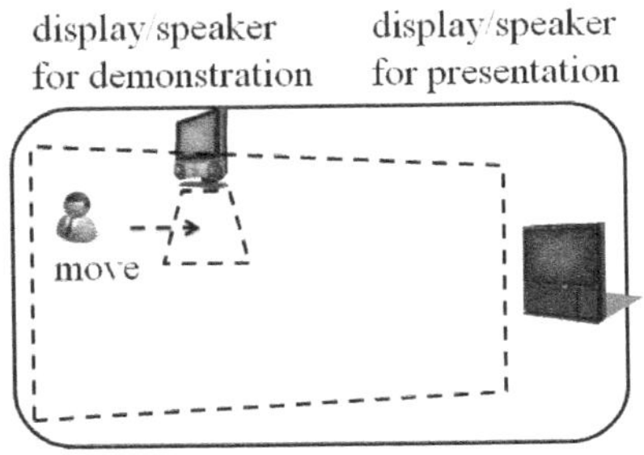

Figure 7.5. *Scope inclusion relationship in case study*

B-a1. When a user requests to start supporting a presentation, the largest display dis_p and the speaker system sp_p are turned on with the presentation contents.
B-a2. When a user requests to stops the presentation dis_p and sp_p are turned off.

B-b1. When a user comes near the display and the speaker for demonstration (dis_d and sp_d), the devices are turned on and the demonstration is started.
B-b2. When a user goes away from dis_d and sp_d, the devices are turned off and the demonstration is stopped.

Figure 7.6. *Initial behavior specification in case study*

The initial behavior of the system is illustrated in Figure 7.6. The display and speaker for presentation are activated by a user's request, whereas those for demonstration are activated by the user movement. The figure only gives an informal description because of the space limitation.

7.4.2. *Analyzing sound conflicts*

7.4.2.1. *Initial specification*

Suppose the initial specification is verified against the property regarding strong prevention of sound conflicts, which is provided by the framework, i.e. "sounds of two different speakers are never received by a user at the same time". This property does not hold because a presentation can start when a demonstration is running. This fact can be easily detected by model checking with one user.

Although the initial specification seems too naive, it should be noted that the proposed framework allows for starting with running or verifying this simple one, because it just enumerates all the possibilities that can occur under the given specification; not requiring a "full" specification.

7.4.2.2. *First modification*

Given the result, developers may add the following rule.

> B-b3. When the presentation is started and if the display and speaker are on, turn both of them off.

This behavior is based on a higher priority of the presentation than that of the demonstration.

This modification turns out to be insufficient after the next run of model checking, because this behavior is only for a situation in which the presentation starts when the demonstration is running.

7.4.2.3. *Second modification*

To handle the other situation in which the demonstration starts when the presentation is running, **B-b1** is modified as follows:

> **B-b1**. When a user comes near dis_d and sp_d, *and if the presentation is not running*, the devices are turned on and the demonstration is started.

Even with the second modification, the property does not hold for strong prevention of sound conflicts. This is because the ECA rule corresponding to B-b3 has a delay, and there is an intermediate state in which a user hears sounds from both the presentation and demonstration.

7.4.2.4. *Third modification*

A possible modification is to integrate the ECA rule of B-b3 into that of B-a1 so that the action to stop the demonstration is taken together with that to start the presentation.

B-a1 When a user requests to start supporting a presentation, the largest dis_p and the sp_p are turned on with the presentation content, *and dis_d and sp_d are stopped if they are on*.

Removal B-b3 is removed.

This modification finally leads to satisfaction of the property.

Another approach is to use a property for weak prevention instead of one for strong prevention to accept undesirable situations if they are handled and stopped. In this case, it is nonsense if the sound conflict is stopped eventually by a user leaving the scope. The setting of a user that remains doing nothing can then be used for meaningful verification.

7.4.2.5. *Solution by environmental assumption*

If the assumption or requirement can be changed so that the scopes for the presentation and demonstration are disjoint, it will be a much simpler solution. In that case, it is still useful to clarify such an assumption, which is often implicit, to enable rigorous understanding and analysis.

7.4.3. *Further scenarios*

The scenario discussed above is simple as a specification, even though it showed that an intuitive but imprecise specification can include some deficiencies if modeled and checked formally.

Other scenarios have also been explored, as briefly described below.

Multiple Users and Demonstration Devices: multiple users and devices are considered to be more realistic.

Mobile Devices: use of mobile devices is considered as an option to see or hear the content of the demonstration, when the user wants to do so or cannot enter either the visual scope or audio scope for some reason (e.g. too crowded). One of the verified properties is that a user can receive both the visual and audio content using either the embedded or mobile device. This property is weak because the user can suddenly move and leave one of the visual or audio scopes.

Content Delivery in Any Media: two types of content are delivered by dis_d and sp_d. However, it may be possible to change the requirement so that the two types of content are delivered in one of the two devices, e.g. using a subtitle on the display instead of oral explanation from the speaker. One of the verified properties, therefore, is that a user can receive both types of content (regardless of the media or the devices). Additional scenarios use this modified requirement to test situations in which the demonstration can still run in mute mode when the presentation is running, as well as when the user is hard of hearing.

7.5. Related work

Analysis of system behaviors described in ECA rules has been considered essential in the areas of ubiquitous computing and pervasive computing. A study [SHA 06] discussed conflicts or incompleteness of ECA rules. In this chapter, we considered a wider space of states including environments and users, as well as an instantiation as a model checking framework.

In the areas of mobile and ubiquitous computing, it has been investigated how to model physical aspects in contexts [STR 04]. Much effort has been made to support design, implementation and execution aspects, e.g. middleware [RAN 03]. Regarding the formal modeling of physical aspects, hierarchical structures have been widely used to model physical scopes or locations such as rooms and buildings. Ambient calculus is a well-known formalism for such hierarchical structures, in the form of process algebra [CAR 98]. Studies (e.g. [RAN 08]) discussed the use of such formalisms for analysis of location-dependent application behavior. In this chapter, we considered an extended model that includes virtual scopes for physical interactions; thus, it does not limit the structure of scopes to a tree. In addition, tool implementation was not often provided in these studies on theoretical process algebra.

Modeling of virtual scopes for effective physical interactions was originally proposed in [SAT 07]. However, that study was focused on the implementation of event-handling mechanisms at runtime, not the development phase and tasks involved. The study also did not take into account scope overlap, as it applied tree-based modeling.

Use of scope-based modeling with awareness of scope overlaps was discussed for formal verification of context-aware algorithms [SCH 09]. However, it targets verification of an algorithm for a certain type of computation, not the whole design of smart spaces.

Ontology-based modeling has been used in ubiquitous computing [RAN 03, CHE 03]. It potentially allows for modeling of scopes discussed in this chapter, as it only uses resource description framework (RDF) or other frameworks to describe arbitrary concepts and their relationships. However, no abstract modeling approach for physical interactions has been discussed, as well as the use of models for verification of smart space design. More uses of semantics or ontology will facilitate linking these studies with the presented framework.

The proposed framework is an extension of a previous study by some of the authors [ISH 09], which did not discuss a concrete tool implementation or specific features for verification. There is also an ongoing study to extend the proposed framework into runtime configuration by end users [YAG 15]. In the study, model checking is similarly used but to present different conflict scenarios to the end users so that they can imagine the situation and set priorities properly.

7.6. Concluding remarks

In this chapter, we have discussed the application of formal verification techniques, specifically model checking, to smart space applications that intensively use actuators, thus, affect users and environments. Specifically, we proposed a framework to model and verify the essences of physical interaction between users and devices. We believe the presented approach provides the essential foundation for dependability of smart space applications that will have increasingly significant roles in human and social activities.

7.7. Acknowledgments

This work is partially supported by the EU-FP7/JP-NICT ClouT Project (608641). The authors thank Kayoko Yamamoto, Basem Suleiman, Marvin Ceschel and Miki Yagita for discussions and implementations of the presented work.

7.8. Bibliography

[BER 01] BERARD B., BIDOIT M., FINKEL A. *et al.*, *Systems and Software Verification: Model-Checking Techniques and Tools*, Springer, 2001.

[CAN 14] CANO J., RUTTEN E., DELAVAL G. *et al.*, "ECA Rules for IoT Environment: A Case Study in Safe Design", *Workshop on Quality Assurance for Self-Adaptive, Self-Organising Systems (QA4SASO 2014) at SASO 2014*, pp. 116–121, September 2014.

[CAR 98] CARDELLI L., GORDON A.D., "Mobile Ambients", *Foundations of Software Science and Computation Structures: First International Conference*, pp. 140–155, 1998.

[CHE 03] CHEN H., FININ T., JOSHI A., "An Intelligent Broker for Context-Aware Systems", *The 5th International Conference on Ubiquitous Computing (Ubicomp 2003)*, pp. 183–184, 2003.

[DEY 01] DEY A.K., "Understanding and Using Context", *Personal and Ubiquitous Computing*, vol. 5, no. 1, pp. 4–7, 2001.

[DIT 95] DITTRICH K.R., GATZIU S., GEPPERT A., "The Active Database Management System Manifesto: A Rulebase of ADBMS Features", *The 2nd Workshop on Rules in Database Systems (RIDS'95)*, pp. 1–17, 1995.

[HEN 06] HENRICKSEN K., INDULSKA J., "Developing context-aware pervasive computing applications: Models and approach", *Pervasive and Mobile Computing*, vol. 2, no. 1, pp. 37–64, February 2006.

[ISH 09] ISHIKAWA F., SULEIMAN B., YAMAMOTO K. *et al.*, "Physical Interaction in Pervasive Computing: Formal Modeling, Analysis and Verification", *The 2009 International Conference on Pervasive Services (ICPS'09)*, pp. 133–140, 2009.

[MUN 14] MUNIR S., STANKOVIC J.A., "DepSys: Dependency aware integration of cyber-physical systems for smart homes", *ACM/IEEE 5th International Conference on Cyber-Physical Systems (ICCPS 2014)*, pp. 127–138, April 2014.

[RAN 03] RANGANATHAN A., CAMPBELL R.H., "A Middleware for Context-Aware Agents in Ubiquitous Computing Environments", *ACM/IFIP/USENIX International Middleware Conference 2003*, pp. 143–161, June 2003.

[RAN 08] RANGANATHAN A., CAMPBELL R.H., "Provably Correct Pervasive Computing Environments", *The 5th IEEE International Conference on Pervasive Computing and Communications (PerCom'08)*, pp. 160–169, 2008.

[SAT 01] SATYANARAYANAN M., "Pervasive Computing: Vision and Challenges", *IEEE Personal Communications*, vol. 8, pp. 10–17, August 2001.

[SAT 07] SATOH I., "A Location Model for Smart Environments", *Pervasive and Mobile Computing*, vol. 3, pp. 158–179, 2007.

[SCH 09] SCHMIDTKE H.R., WOO W., "Towards Ontology-Based Formal Verification Methods for Context Aware Systems", *The 7th International Conference on Pervasive Computing (Pervasive 2009)*, pp. 309–326, 2009.

[SHA 06] SHANKAR C., CAMPBELL R., "Ordering Management Actions in Pervasive Systems using Specification-enhanced Policies", *The 4th IEEE International Conference on Pervasive Compputing and Communications*, pp. 234–238, 2006.

[SPE 15] SPEC PATTERNS, "About Specificaiton Patterns", http://patterns.projects.cis.ksu.edu/, 2015.

[SPI 15] SPIN ROOT, "SPIN – Formal Verification", http://spinroot.com/, 2015.

[STR 04] STRANG T., LINNHOFF-POPIEN C., "A Context Modeling Survey", *Workshop on Advanced Context Modelling, Reasoning and Management, UbiComp*, 2004.

[TWI 09] TWIDLE K., DULAY N., LUPU E. *et al.*, "Ponder2: A Policy System for Autonomous Pervasive Environments", *The 5th International Conference on Autonomic and Autonomous Systems (ICAS 2009)*, pp. 330–335, April 2009.

[WEI 99] WEISER M., "The Computer for the 21st Century", *SIGMOBILE Mobile Computing and Communications Review*, vol. 3, no. 3, pp. 3–11, ACM, 1999.

[YAG 15] YAGITA M., ISHIKAWA F., HONIDEN S., "An Application Conflict Detection and Resolution System for Smart Homes", *International Workshop on Software Engineering for Smart Cyber-Physical Systems (SEsCPS 2015)*, pp. 33–39, May 2015.

8

SmartSantander: A Massive Self-Managed, Scalable and Interconnected IoT Deployment

The concept of smart city [HER 11] is about improving the efficiency of city services and facilitating the sustainable development of cities, progressing toward smart urban environments and engaging populations (living in these cities) in this endeavor. To foster user support, involvement and engagement in ICT-based innovation areas, smart cities are thought to be excellent playgrounds for Future Internet (FI) research and experimentation. For this purpose, under SmartSantander FP7 project framework, a massive IoT deployment was carried out in the city of Santander (in the north of Spain). This deployment is supported by a novel architecture, also designed and implemented during the project, which covers a threefold approach: service provision, network management and experimentation support. Both deployment and architecture are characterized by a high degree of scalability and flexibility, thus allowing for the addition of new physical devices as well as the improvement of developed software modules. In this sense, within the ClouT EU-JP FP7 project, the aforementioned deployment and infrastructure will serve as basis for the development of new use cases in Santander, thus enriching retrieved information with new data sources, as well as generating new entities or improving existing ones to achieve a better network performance.

8.1. Introduction

Under the SmartSantander project framework, an experimental test facility for the research and experimentation of architectures was used, enabling key technologies, services and applications for the Internet of Things in the context of a large city (in this case, the city of Santander located in the north of Spain). This facility provides the European research community with a one-of-a-kind platform,

---

Chapter written by José Antonio GALACHE, Juan Ramón SANTANA and Luis MUÑOZ.

suitable for large scale experimentation and evaluation of IoT concepts under real-life conditions.

The SmartSantander project provides a two-fold opportunity. On the one hand, the research community benefits from using such a unique infrastructure which allows for true field experiments, providing researchers with the capacity to reserve specific resources within the whole network and for a determined time period within which to run their experiments. On the other hand, different services fitting citizens' requirements and associated with public services have been also used.

To manage this efficiently, a flexible and scalable architecture was defined and developed, allowing for the addition of new hardware devices and the improvement of software entities that compose it, under the framework of other research projects (e.g. ClouT project), and the improvement of the city's public services.

8.2. SmartSantander: novel architecture for service provision and experimentation

As previously mentioned, the two-fold (experimentation + service) approach of the project translates into a reference model architecture for the infrastructure used, not solely relying on particular service-specific deployments, but providing service providers and technology manufacturers with a testbed-like urban experimentation environment, where early-day smart cities will greatly benefit from the capability of experimenting with the development of future services.

Figure 8.1 shows the aforementioned architecture [SAN 14] as a three-tiered network approach: IoT node tier, gateway (GW) tier and testbed server tier.

The *IoT node tier* comprises the majority of the devices used in the testbed infrastructure. It is composed of diverse heterogeneous devices, including miscellaneous sensor platforms, tailor-made devices for specific services as well as Radio-Frequency Identification (RFID) and Near Field Communications (NFC) tags. These devices are typically resource-constrained and host a range of sensors and, in some cases, actuators. Other devices such as mobile phones and purpose-built devices with reasonable computing power (e.g. mobile devices in vehicles), as well as providing wide area communication capabilities, behave as IoT nodes in terms of sensing capabilities and as GW nodes regarding processing and communication capabilities.

The *GW tier* links the IoT devices on the edges of the capillary network to the core network infrastructure. IoT nodes are grouped in clusters that depend on a GW device. These nodes gather and process information retrieved locally by IoT devices within its cluster. The GW device manages (transmission/reception of commands) them, thus scaling and easing the management of the entire network. GW-tier devices are typically more powerful than IoT nodes in terms of memory and processing capabilities, also providing faster and more robust communication interfaces. GW devices allow for the virtualization of IoT devices. This enables the instantiation of emulated sensors or actuators that behave in all respects similar to the actual devices.

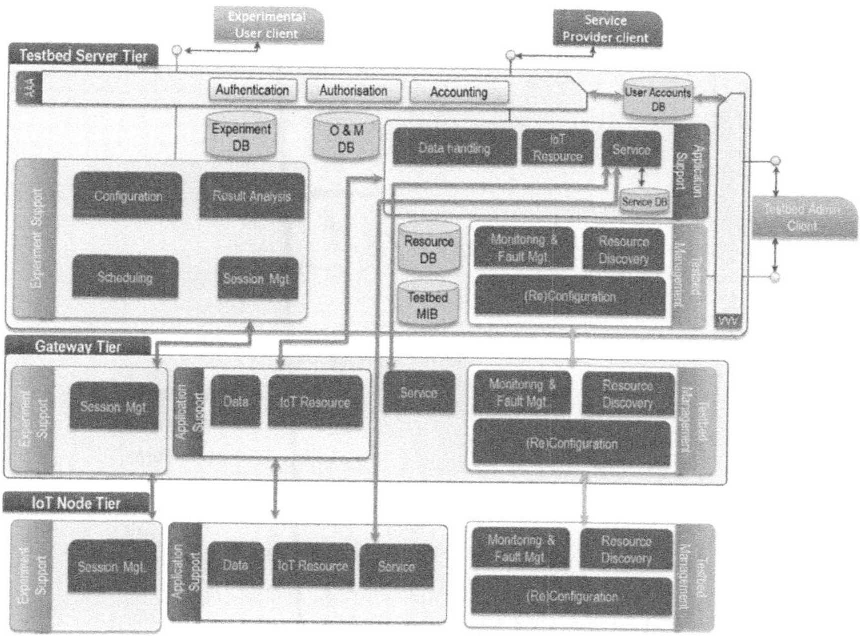

Figure 8.1. *Platform high-level architecture and building blocks*

The *test bed server tier* provides powerful computing platforms with high availability and is directly connected to the core network. The servers are used to host IoT data repositories and application servers. Server-tier devices receive data from all GW-tier nodes. As a final step, the concept of federation is supported by the architecture. Servers managing networks located in different physical locations can connect among themselves to allow users of the platforms to transparently access IoT nodes that are used in different testbeds.

Figure 8.2 shows the low-level architecture [GAL 13a] that maps the functionalities and services previously described onto specific building blocks. From the user perspective, there exists a user directory with the corresponding users accessing to the platform, being able to access to the retrieved values in an asynchronous way or through the corresponding data repository (i.e. FIWARE).

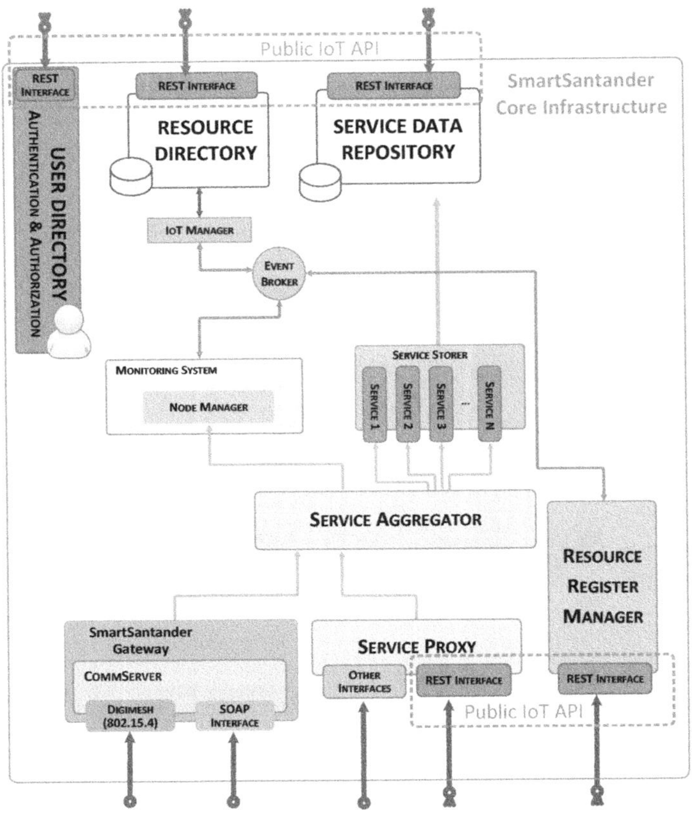

Figure 8.2. *Low-level platform architecture and implemented modules*

To access and manage IoT devices, the SmartSantander Gateway is in charge of the IoT devices installed during the lifetime of the project, thus also allowing to experiment on top of these IoT devices by sending/receiving commands to/from them, through the Digimesh interface (on top of IEEE802.15.4) and the SOAP interface (on top of GPRS). To add new devices, they can be added to the platform through the aforementioned gateway if using Digimesh/GPRS interfaces or directly using the Public IoTAPI (to be further described in the next sections) which offers a REST interface for injecting information from external devices to the platform through service proxy. In the case of these external devices, they should be previously registered using the resource register manager, which will use the event broker to update the resource directory where all resources are registered. All the registered nodes will be managed by the node manager, in charge of monitoring information generated by the platform gathered and sent by the general aggregate manager and service aggregator, thus keeping their status in the resource directory updated. Additionally, values retrieved by the registered nodes are managed by the service aggregator and service proxy (a dedicated instance for each of the specific services addressed), injecting data into the corresponding service data repository. The public IoT manager will also be managing platform access for any additional users, providing an authentication method based on the use of API keys that must be included in the headers of every http request made. API keys are randomly generated and assigned to different users depending on their needs, thus generating X.509-based certificates. The use of API keys allows network managers to ban users from accessing the platform in the case that a misuse of the API is detected.

8.3. SmartSantander deployment: use cases

The architecture shown in the previous section is in charge of managing the Santander testbed [GAL 13a], which is composed of approximately 3,000 IEEE 802.15.4 devices, 200 devices, including GPS/GPRS capabilities, and 2,000 joint RFID tag/QR code labels used both at static locations (streetlamps, facades, bus stops) and on-board public vehicles (buses, taxis). Its deployment (shown in Figure 8.3) consists of the development of different use cases within the project, as described below:

– *Static Environmental Monitoring*: Approximately 2,000 IoT devices, installed (mainly in the city center) on streetlamps and facades, are equipped with different sensors which offer measurements on different environmental parameters, such as temperature, CO levels, noise and luminosity. All these devices are equipped with two independent IEEE 802.15.4 modules, one running the Digimesh protocol (proprietary routing protocol) intended for service provision (environmental measurements) as well as network management data transmission, and another (that implements a native 802.15.4 interface) associated with data retrieved from experimentation issues.

– *Mobile Environmental Monitoring*: To extend the aforementioned static environmental monitoring use case, apart from measuring parameters at static points, 150 devices located on public vehicles (buses, taxis) retrieve environmental parameters associated with determined parts of the city. Modules installed in these vehicles are composed of a local processing unit in charge of sending (through a GPRS interface) the values (geolocated) retrieved by both the sensor board and the CAN-Bus module. The sensor board measures different environmental parameters, such as CO, NO_2, O_3, particulate matters, temperature and humidity, whereas the

CAN-Bus module retrieves and records the main parameters associated with the vehicle, such as position, altitude, speed, course and odometer. Furthermore, an additional 802.15.4 interface is also included to carry out experimentation, interacting with the aforementioned static devices, so-called vehicle-to-infrastructure (V2I) communication.

– *Parks and gardens irrigation*: Approximately 50 devices have been used in two green zones of the city to monitor irrigation-related parameters, such as moisture, temperature pressure and humidity, to make irrigation as efficient as possible. In terms of processing and communication issues, these nodes are the same as those used for static environmental monitoring, implementing two independent IEEE802.15.4 communication interfaces.

– *Outdoor parking area management*: Almost 400 parking sensors (based on ferromagnetic technology), buried under the asphalt, have been installed at main parking areas of the city center, to detect parking availability in these zones.

– *Guidance to free parking lots*: Taking information retrieved by the used parking sensors, 10 panels located at the main streets' intersections have been installed to guide drivers toward the available parking lots.

– *Traffic Intensity Monitoring*: Approximately 60 devices located at the main entrances to Santander have been used to measure main traffic parameters, such as traffic volumes, road occupancy, vehicle speed or queue length.

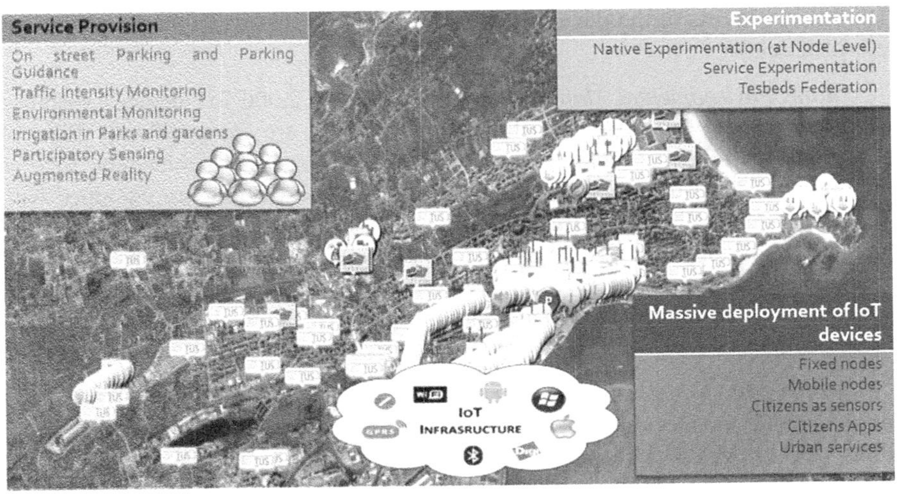

Figure 8.3. *Santander city deployment*

All of these use cases provide a different service and offer the retrieved data to users, so-called experimentation at service level. Static and mobile environmental monitoring and parks and gardens irrigation also offer the possibility of carrying out experimentation at the node level, offering an additional communication interface.

Apart from the aforementioned use cases, two citizen-oriented services have been used, thus including corresponding applications for Android and IOS operating systems, to foster the citizens' involvement.

– *Augmented Reality*: This service includes information about more than 2,700 places in the city of Santander, classified in different categories: beaches, parks and gardens, monuments and shops. To complement and enrich this service, 2,000 RFID tags/QR code labels have been used, offering the possibility of "tagging" points of interest (POI) in the city such as touristic POI, shops and public places (parks, squares). On a small scale, the service provides the opportunity to distribute information in the urban environment as location-based information.

– *Participatory Sensing*: In this scenario, users utilize their mobile phones to send physical information to the SmartSantander platform in an anonymous way, e.g. GPS coordinates, compass, environmental data such as noise, temperature. Users can also subscribe to services such as "the pace of the city", where they can get alerts for specific types of events currently occurring in the city. Users can themselves also report the occurrence of such events, which will subsequently be propagated to other users that are subscribed to the corresponding types of events.

It is important to highlight that, in the same way as the aforementioned use cases, information retrieved by these two services is made available to the SmartSantander platform for other users to experiment with (i.e. experimentation at service level).

8.4. SmartSantander interacting with ClouT

As previously mentioned (see Chapter 5), the ClouT project takes advantage of SmartSantander by using data retrieved by it but also by inheriting management tools, operating modules, communication interfaces and description standards; all of them are intended for addressing IoT device naming, description and management, as well as storage, processing and sending of information associated with a massive IoT deployment, thus connecting these functionalities with those provided by the ClouT project.

8.4.1. *IoT device naming*

Based on the experience of the SmartSantander project, and with the help of some of the considerations indicated in the RFC2141 and RFC3406 compliances, a set of rules for building the corresponding IoT device identifier have been defined.

First of all, regarding to the naming and structure, some restrictions should be considered:

– valid characters: ASCII table and characters ".", "-" and "_". Character ":" is used as levels separator;

– no case sensitive: recommended to use a uniform resource name (URN) in lower-case characters;

– length: in principle bounded to 255 characters.

The URN will be composed of four parts: prefix, domain, provider_index and local_id:

– *Prefix*: A common identifier for all the resources. In this sense, just the "urn:" prefix will be used, indicating that it is a unique identifier. Additionally, for indicating an IoT node used in an experimental way, the "x-iot" prefix could be used. Nevertheless, considering that some of the nodes used in the facility support different public services, the use of this prefix may not be appropriate for all the devices, so its use is optional.

– *Domain*: Associated with each of the different instances of the project. In this sense, the domain could make reference to the project and the different cities/partners within the consortium. Something similar to this "clout:fujisawa", when referring to a resource located in Fujisawa.

– *Provider/Manufacturer/Owner index or associated service*: In this case, two options may be considered. In the first, the identifier would be associated with the company that provides the service, the manufacturer of the devices or the owner of the facility, being assigned a random alphanumeric or a sequential numeric value to each of the providers. In the second option, the identifier will be assigned a description of the provided service, with which the node is associated. This translates to the generated identifier being clear in terms of the service (environment, transport, etc) to that node is feeding with its measurements, but not so flexible in terms of including information of the manufacturer/provider.

– *Local ID*: Proprietary ID established by the owner. Its format and content are flexible, but must respect the structure restrictions previously defined.

Some examples of naming are indicated below:

"urn:[x-iot]:clout:santander:38bf2t:temp10 (1st option)

"urn:[x-iot]:clout:santander:environment:temp10 (2nd option)

The examples define a node installed in the city of Santander, indicating the corresponding manufacturer as a random alphanumeric identifier in the first option or the associated service (environment) in the second option. Finally, the local id is temp10, also used as the proprietary id by the corresponding manufacturer.

8.4.2. *IoT device description*

Apart from the unique identifier associated with the node, the IoT device must also store information related to the node description, which will be stored in the Orion Context Broker. For this purpose, this entity is comprised of a multi-level JSON structure and a tuple (name, type and value) for storing attributes and metadata associated with a specific node:

– Time Installation: The time a node has been installed (real nodes), added to the platform (legacy device) or created (virtual node).

– Description: Nodes can be classified according to their mobility: fixed, mobile, semi-mobile, or their abstraction status: real, sensorized or virtual. In this sense, semi-mobile nodes could be defined as nodes that are taking measurements during a specific amount of time in a fixed location, but that can be moved to another location to keep on taking measurements at this new fixed position during a specific period of time. This also includes the main characteristics of the device in terms of sensing capabilities, processing and memory capacity, estimated battery life, name of the manufacturer, communication interfaces and so on. It could also include additional functionalities associated with the node, in terms of remotely sending/receiving commands or capacity for remote firmware update.

– Management: Shows the operational status of the node, indicating if it is working properly, being replaced, turned off or other specific status that could be defined and added.

– Position: GPS coordinates of the device. This is only applicable for fixed nodes; for mobile and semi-mobile nodes; position is associated with each of the measurement.

– Capabilities: Number of measuring attributes associated with a determined node. For each of these capabilities, they will be indicated the following parameters:

- name: identifier of the capability according to a specific dictionary that assures the univocal identification of a determined capability, independent of the device location where it is being measured (i.e. temperature as the common identifier for ambient temperature measured by a node);

- type: this field describes the specific phenomenon that is specifically measured;

- value: it indicates the value of the measurement carried out;

- metadata: different types of metadata could be included, all of them compliant with the tuple (name, type, value) structure. For each of the capabilities, one of the metadata will be associated with the unit of measurement of the physical parameter measured by the node, and the other to the status of the node, which is related with the capacity of measurement of a determined variable.

– Log attribute: specifies changes in the node, such as location, new sensing capabilities or sensor status, among others. In this sense, it would include two metadata, one containing the timestamp of the modification carried out and the other with the description of the change being carried out. It is important to highlight that an entry in the log attribute will translate in the modification of some of the attributes previously mentioned, in the way of:

- if the node has changed its location, then it will consist of the updated latitude and longitude coordinates;

- if new sensing capabilities are added, then it will be included as a new attribute with the corresponding unit and status metadata;

- if the status of some of the sensors has changed, it will be updated accordingly. In this sense, as will be further explained in the next sections, this status could take five different values, defined by the self-healing tool [NGU 15].

Figure 8.4 shows an example of the generic structure of the resource description.

```
"contextElements": [{
   "type": "urn:clout:entityType:iotdevice",
   "isPattern": "false",
   "id": " urn:clout:santander:environment:temp10",
   "attributes": [{
      "name": "TimeInstallation",
      "type": "urn:clout:datatype:ISO8601",
      "value": "2015-10-29T10:46Z"
   }, {
      "name": "Description",
      "type": "urn:clout:datatype:string",
      "value": "Acoustic sensor"
```

```
}, {
  "name": "Management",
  "type": "urn:clout:datatype:ISO8601",
  "value": "2015-11-29T10:46Z",
  "metadatas": [{
    "name": "status",
    "type": "string",
    "value": "READY"
  }]
}, {
  "name": "position",
  "type": "fixed",
  "value": "40.2344,-3.87768",
  "metadatas": [{
    "name": "location",
    "type": "string",
    "value": "WGS84"
  }]
}, {
  "name": "batteryLevel",
  "type": "urn:clout:attributeType:battery",
  "value": "76",
  "metadatas": [{
    "name": "unit",
    "type": "urn:clout:uom:integer",
    "value": "battery percentage"
  },{
    "name": "status",
    "type": "urn:clout:datatype:string",
    "value": "FAILURE"
  }
  ]
},{
```

```
        "name": "temperature",
        "type": "urn:clout:attributeType:temperature:ambient",
        "value": "23.5",
        "metadatas": [{
          "name": "unit",
          "type": "urn:clout:uom:float",
          "value": "degreeCelsius"
        },{
          "name": "status",
          "type": "urn:clout:datatype:string",
          "value": "READY"
        }
        ]
      }, {
        "name": "Updates_log",
        "type": "urn:clout:datatype:string",
        "value": "Log of changes: position, new sensor cappabilities added, .. It will be
created an empty with the date and brief explanation of the modification carried out",
        "metadatas": [{
          "name": "TimeUpdate1",
          "type": "urn:clout:datatype:ISO8601",
          "value": "2015-10-29T10:46Z"
        },{
          "name": "LogUpdate1",
          "type": "urn:clout:datatype:string",
          "value": "Change of location"
        }
        ]
      }]
    }],
    "updateAction": "APPEND"
}
```


Figure 8.4. *Resource description*

As can be observed in Figure 8.4, the described node is a static device located in Santander that measures two environmental values: temperature and noise, and with a change of location log message in its description.

8.4.3. *IoT resource manager*

As was indicated in the previous section, a specific field with a log of the different changes in the node attributes in included in the node description. In this sense, the node manager entity, previously in charge of updating node status, will be fed by accurate information generated by a data-healing component in the ClouT platform [NGU 15] in terms of the node attributes' status. Figure 8.5 shows the corresponding interaction.

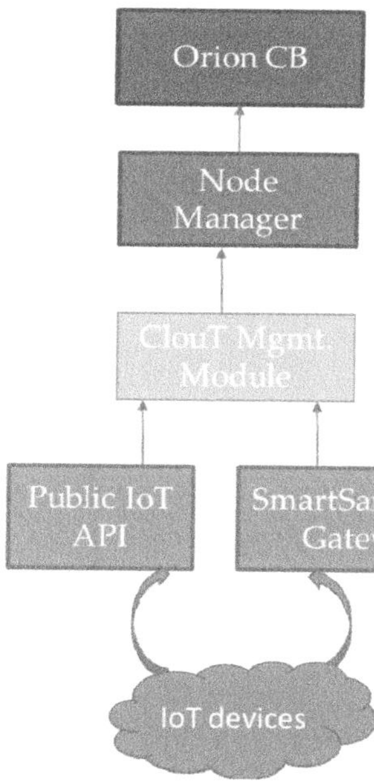

Figure 8.5. *SmartSantander resource manager architecture*

As can be seen from Figure 8.5, IoT devices are managed by both the SmartSantander Gateway and Public IoT API, thus offering information through the ClouT platform to the corresponding ClouT management module. This module is in charge of processing the information received from each device, thus determining the status of the node and each of its attributes, according to values such as the time interval among measurements or the retrieved value for each of the attributes. The node manager will be fed this information, after which it will send the corresponding update command to the Orion Context Broker, indicating the values to be modified as well as the entry to be added to the Log attribute. This way, the resource description will be kept up to date, with the log showing any changes made to the node over time.

8.4.4. *Virtualization module*

In both the SmartSantander gateway and public IoT API levels, the virtualization module can be defined as an extension of the basic capabilities offered by subjacent resources, offering four different types of virtualization:

– resource combination: virtualized resource in a repository;

– resource abstraction: new combined sensors that could be offered to other users (with the possibility of being also virtualized in the repository);

– mobile sensor: mobile phones or tablets can be virtualized, such that citizens/users will behave as virtual sensors (with the possibility of being also virtualized in the repository);

– sensor history: information retrieved, by both real and virtual resources, can be aggregated to the historical data already stored.

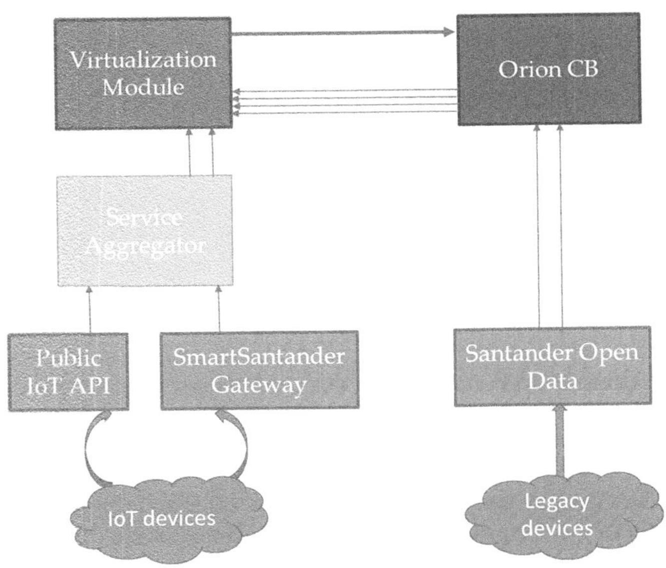

Figure 8.6. *Virtualization architecture*

As can be seen from Figure 8.6, information from both IoT and legacy devices in the city of Santander are used as data sources, thus using the Public IoT API and the SmartSantander GW entities and the Santander Open Data engine for retrieving data from IoT and legacy devices, respectively. Information from IoT devices will be gathered by the service aggregator, in charge of loading information into the virtualization module. Regarding data coming from legacy IoT devices, this will be directly loaded onto the Orion Context Broker, which will be in charge of providing this information to the virtualization tool. This way, once a new virtual node needs to be generated, the virtualization module will ask for the information associated with the set of legacy and/or IoT devices that will be part of the virtual node. The virtualization module will carry out the corresponding operation (sum, average, median, etc.) over the selected nodes, thus creating a new virtual node, whose description as well as the data generated by it will be stored in the Orion CB, thus behaving as an additional node included in the portfolio of nodes offered by the Orion CB to the additional users. Orion CB support, as well as the Santander open data portal as additional data source, represents the main updates carried out in this third year.

To carry out the creation of a new virtual device, a web application has been developed which allows users to easily access the information provided by a set of used nodes and combine these accordingly to generate a virtual device, as shown in Figure 8.7.

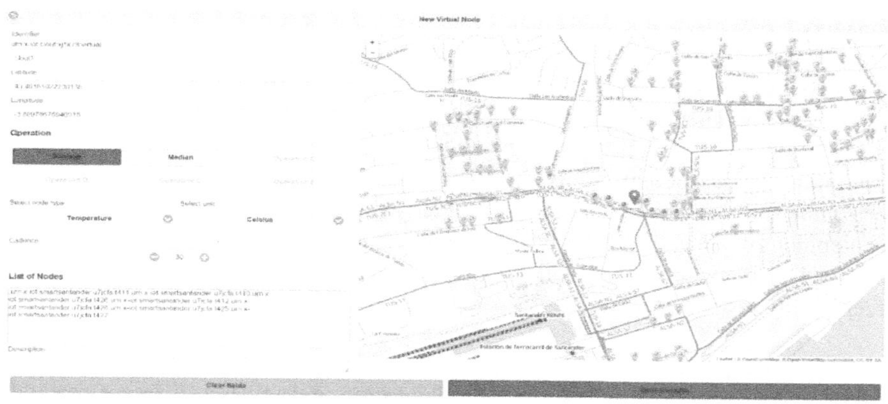

Figure 8.7. *Virtualization web application*

As can be seen from Figure 8.7, a device is first of all selected as an identifier for the node, where the prefix name is already defined (see section 8.4.1) and where a user can add a suffix for identifying the node. After, the latitude and longitude associated with the virtual node are indicated, which can be automatically generated by clicking on the map directly. Next, the type of operation to be carried out over the set of nodes that are part of the virtual node is selected, indicating the type of nodes (according to the attribute to be measured), as well as the cadence of the measurement (expressed in seconds). Finally, in the last section, the set of nodes that compose the virtual node are selected by clicking on them in the map (red nodes are the ones that have been selected) or by typing the name in the corresponding text box. Once finalized, at the bottom part of the screen, there are two buttons, one for saving changes and creating the virtual node (on the right-hand side), and another one to clear all the fields and start again.

It is important to highlight that the creation of new virtual resources and, consequently, new services associated with either these new resources or a combination of them, offers the possibility of supporting a great portfolio of functionalities.

8.5. Conclusions

The smartcity scenario stands at the meeting point between the provision of a facility for offering different types of services and the deployment of a massive testbed on which to run a great variety of experiments. This translates to a wide portfolio of stakeholders, ranging from citizens to researchers, including public administration and service providers, all under the supervision of the network managers.

To cover the requirements associated with all the aforementioned stakeholders, a city-scale experimental research facility in support of typical applications and services for a smart city, the first of its kind, was used in the SmartSantander project. This unique experimental facility is sufficiently large, open and flexible to enable horizontal and vertical federation with other experimental facilities and stimulates development of new applications by users of various types, including experimental advanced research on IoT technologies as well as realistic assessment of users' acceptability tests.

The ClouT project takes advantage of the openness and flexibility of the used testbed at hardware and software levels, inheriting and reusing different modules, interfaces and protocols developed within the SmartSantander project, mainly related to the addition, registration and management of new IoT devices. In terms of management, accurate information from the ClouT platform is used to keep node statuses up to date within the Orion Context Broker, the entity where all resource context information will be stored.

Considering the updated information provided within ClouT framework, a virtualization tool has been developed to generate new virtual devices that support purpose-specific applications.

Finally, the improvement of the SmartSantander platform, under the umbrella of ClouT project, in terms of both management capabilities and number of data sources provided, clearly illustrates the capability of the SmartSantander platform as a base architecture on which to support and integrate new functionalities.

8.6. Bibliography

[GAL 13] GALACHE J.A., GUTIÉRREZ V., SANTANA J.R. *et al.*, "SmartSantander: a joint service provision facility and experimentation-oriented testbed, within a smart city environment", *Future Network & Mobile Summit*, Lisbon, Portugal, July 2013.

[HER 11] HERNÁNDEZ-MUÑOZ J.M., VERCHER J.B., MUÑOZ L. *et al.*, "Smart cities at the forefront of the future internet", in DOMINGUE J. *et al.* (eds), *Future Internet Assembly*, Springer-Verlag, Berlin, Heidelberg, 2011.

[NGU 15] NGUYEN T.A., AIELLO M., YONEZAWA T. *et al.*, "A self-healing framework for online sensor data", *Computer Networks: 3rd International Workshop on Self-Aware Internet of Things (Self-IoT 2015)*, pp. 295–300, July 2015.

[SAN 14] SANCHEZ L., MUÑOZ L., GALACHE J.A. *et al.*, "SmartSantander: IoT experimentation over a smart city testbed", *Computer Networks: The International Journal of Computer and Telecommunications Networking*, vol. 61, pp. 217–238, March 2014.

9

Using Context-aware Multi-agent Systems for Robust Smart City Infrastructure

Among the central concerns in the deployment of a smart city infrastructure are the issues of robustness, dependability and privacy-awareness. These issues influence the experience of the user with smart city applications and are strongly related to the large scale of a Smart City. We have developed a set of core requirements for Smart City infrastructures, and we have devised several elements that can be part of an adequate solution. Among these are the distribution of the system, the autonomy of entities, the locality of their behavior and a set of organizational aspects for all entities in the system.

9.1. Introduction

In a smart city [CAR 09, GIF 07] individuals use various personal, shared or public items, such as smart electronic devices or smart appliances, to improve their everyday life. This ubiquitous electronic environment or ambience should provide the right information or perform the right action, at the right time and in the appropriate context, appearing as intelligent to the user. Moreover, "smart" software should be able to anticipate the user's needs and potentially dangerous situations.

All this should be done while respecting the user's privacy and not offering personal information (PI), or personally identifiable information (PII), to organizations or governments, without the explicit agreement and understanding of the user. Reaching this trade-off is especially difficult when a large number of users

Chapter written by Andrei OLARU, Adina Magda FLOREA and Amal EL FALLAH SEGHROUCHNI.

are in need of context-aware information in a prompt manner and they desire it to be free of charge[1] [TEN 13, RUL 15].

In this chapter, we will focus on our experience of how we can achieve the goal of providing smart city users with fresh, relevant information, promptly, without the users needing to offer personal information in exchange, except when the information is absolutely necessary and the user is fully aware of the transfer [TEN 13].

9.1.1. *Smart cities and ambient intelligence*

Though there are many different definitions and opinions as to what a smart city is and what features it offers, it is clear that in a smart city, Information and Communications Technology (ICT) and technology in general are used to greatly improve the life of its citizens and also enhance their ability to work, relax and help each other [DEA 11]. This improvement is brought about by means of two key elements: communication (exchange of information) and artificial intelligence (information processing). More precisely, relevant information must get to where it is needed, and, by using that information, wise decisions must be made and new, higher-level information should be created.

The difference between an ideal smart city and what various smart services offer currently is that in a smart city ICT *pervades all city functions* and areas of interest, and that all services that are available are *connected and unified*. Not only does a smart city feature smart governance, smart energy, smart building, smart mobility, smart infrastructure, smart technology, smart healthcare and smart people [SIN 14], but all of these are also *connected* to form the smart city infrastructure.

Smart cities have many features in common with Ambient Intelligence (or AmI). The goal of AmI is to create an electronic environment that perceives the state and activity of the user to assist the user and improve the user's experience, by processing available information in an intelligent manner [DUC 01]. Similarly, a citizen of the smart city is able to benefit from an electronic environment that is able to assist her/him in daily activities.

The key difference between the two concepts – smart cities and ambient intelligence – is scale. Although AmI can be implemented in a home, a museum or any other type of smaller environment, a smart city is always a large city. Any solution that is proposed for a smart city must work (at least) at this scale [SIN 14, EVA 11]. This means an intelligent system that is able to assist approximately 20 million users and up to three to five times as many devices, if we

1 It is arguable that the great majority of free (as in beer) online services are actually paid for with the users' private information.

only consider personal ones. Added to these numbers are a large number of buildings, vehicles and potentially billions of sensors. Therefore, though it is possible to use some solutions from the field of AmI for the needs of smart cities, it is essential that these solutions scale to the required level.

Another recent concept that is related to smart cities and AmI is the Internet of Things (IoT) [EVA 11]. The IoT is an electronic environment that pervades all appliances, such that all devices and equipment, from light bulbs to notification lights to power plants, are connected to and accessible from the Internet. It is similar at hardware level to the concept of AmI, but it does not mention intelligence. In a way, IoT and AmI technologies are both enablers for smart cities.

One key point in designing infrastructures for smart cities is that the infrastructure must be able to support a wide range of smart applications. Let us take a few examples that we will refer to throughout this chapter:

– data is gathered from traffic lights, traffic cameras and vehicles to determine driving patterns and automatically develop ways in which traffic can be improved. Although all drivers participate in the gathering of information, the specific driving habits of an individual driver cannot be identified;

– a group of people participating in a public event receive information about how to find the local facilities, how to find other participants and how to get to the exit such that all the exits are used equally. Although all participants give information on their location in the building and other indications of their activity, it is not possible for user A to know the location and activity of user B, unless user B allows it explicitly;

– an elderly person goes outside the house and takes a walk through the city. The physician who cares for the person is able to obtain aggregate information about the person's walk. More details on the person's location and whereabouts will become available in the case in which an emergency is detected, but only then; and only to the parties who can help the person.

All of the examples above have some common features: information is obtained about the participants in an activity; this information is used to provide the participants (directly or indirectly) with useful information and actions, in direct relation to their activity (i.e. not for advertising purposes); the information is managed in such a way that it cannot be used to invade the privacy of the individual users. We will discuss privacy-awareness further on the chapter.

9.2. Requirements

To help with the design of the architecture presented in this chapter, we have developed a list of requirements. These requirements are derived from our previous work, especially the tATAmI-2 [BAL 12] and DESIA [MAC 15] architectures. We have grouped these requirements into three categories: context-awareness and integration (CTX-*), scale and dependability (SCL-*) and user's privacy (PRV-*).

These categories reflect the needs that must be satisfied by any smart city software infrastructure: the system must be aware of the user's state and needs, must integrate with other services in use by the user and be able to understand the data that the user or these services provide; the system must be able to scale to the number of users and devices in a large city, remaining dependable and robust; and the system must ensure the privacy of the users and the protection of their personal information.

The list of requirements that we have developed is as follows:

– context-awareness and integration (CTX):

1) the system must be able to retain context information acquired by sensors or created by various components and deliver this information to components that are able to process it further, or to the user,

2) the system must be able to understand what the information refers to and how it is linked to the activity and needs of the user,

3) the system must represent context information in a manner which is easy to understand both by other applications and services and by the user as well,

4) the system must deliver the relevant information to the interested user without any express action or request from the part of the user. The system must be able to anticipate the needed information and deliver it to the user at the appropriate time;

– scale and dependability (SCL):

1) the system must be able to work with a large number of devices and users – in the tens and hundreds of millions,

2) the system must be able to work, even partially, even if not all components are available, i.e. even if a part of the network gets disconnected from the rest, users in that part of the network should still get the information that can get to them (e.g. is generated in the same part of the network),

3) the system must be dependable – in any situation, the user must be able to access the information he/she needs at the time (in reasonable limits), even if completely disconnected from the network and even if the information is not the most recent;

– user privacy and security (PRV):

1) no identifiable information that is generated by the user, or about the user, should be delivered to any other user or service, without the fine-grained permission of the user. The user should be able to control who has access to each piece of information about her/him,

2) the system must be able to aggregate information about large group of users without allowing access to (or ideally, without ever holding) any piece of personal information or information that could be associated with the user later (through correlation with other pieces of information);

One additional, softer, requirement is that the user should be able to use the system in comfort, in what is called *flow*, or optimal experience [CSI 92]. The system should act intelligently and in an anticipative manner such that the user never finds the experience inadequate or frustrating. This is why, ideally, all the needed information must be delivered to the user with no delay, and if possible before the user requests it explicitly, but at the right time and without intrusion. We will call this requirement OPT-1.

Another factor that contributes to an optimal experience is *traceability* – the capability of the system to offer explanations to the user about the state of the system and about what actions were taken proactively and why. The level of complexity of explanations should be incremental, such that users who are more knowledgeable are able to obtain a more detailed explanation. This reassures users and makes the system more user-friendly. We will call this requirement OPT-2.

9.2.1. *The necessary infomation at the right time*

The main target of the architecture that we propose is to deliver to a user in the smart city the right information at the right time, in a reliable manner, without privacy concerns for the user.

However, what is the right information, and when is the right time? We argue that the right information is strongly related to the context of the user at the current time. Context has been defined as "any information that can be used to characterize the situation of entities (i.e. a person, a place or an object) that are considered relevant to the interaction between a user and an application, including the user and the application themselves" [DEY 01].

If we eliminate possible distractions, the user will always need information that is related to his/her current activity, or activities in the user's schedule. If we include distractions, these can come from sources such as the user's spatial context or the user's social context. Therefore, any potentially needed information relates to the user's context in some way, namely with one of the three types of context – activity, space or social [ELF 12].

If we are able to associate a particular piece of information (e.g. a file) with some elements of context (e.g. "file `mytask.doc` is related to Task 2 at my job"), then the system would note that when the user is in that context (e.g. working on Task 2) then the user may need that information. We can recognize the context of the user by following the user's activity (e.g. the user has scheduled some work on Task 2 in his/her personal organizer). This approach is inspired by the work of Brézillon and Brézillon [BRÉ 07], who describe the context as the "dressing" of a focus. That is, when a user has a focus, and we can detect it, we consider all the information relevant that is ascribed to that context, which is related to the focus of the user.

9.2.2. *Robustness, reliability, dependability and trust*

In his seminal work on ubiquitous computing [WEI 95], Weiser recounts that "there is more information available at our fingertips during a walk in the woods than in any computer system, yet people find a walk in the woods relaxing and computers frustrating". Although this is not completely true anymore, there is still a lot of frustration in using computers and "smart" devices in general: relating primarily to two factors the applications are often not smart, either performing repetitive actions uselessly or requiring the user to perform actions repeatedly (i.e. applications are *dumb*); and the lack of network connection or other network issues mean long delays or even the impossibility of accessing the user's own data. Such issues make computing devices not appear to interact *naturally* with the user and they are not seen as dependable, the user cannot rely on them in such cases. The system is not seen as comfortable to use, and the flow is broken.

Ideally, with sufficient knowledge about the user's schedule, a context-aware application should be able to anticipate the user's need for information and cache that information on the user's device. Conversely, when possible, user's data should be backed up on other devices of the user as well, such that if the data on the current device is lost, it is already present on other devices. Although this type of behavior is already present in cloud-based applications, these applications rely on centralized servers which are far away from the user's access. The proposed architecture attempts to eschew centralization by creating a network of the user's devices overlaid onto the network infrastructure available to the user at the time (much like opportunistic routing [CHA 07]).

9.2.3. *Privacy and personal information*

Further reason for not using centralized servers that belong to corporate entities is privacy and the protection of personal information. Recent events and disclosures regarding surveillance show that the massive amounts of data that a user's activity produces are not only stored and processed by companies for targeted advertising, but are also available for inspection by the government (be it the government of the country of which the user is a citizen of, or governments of countries the user has no relation to), potentially leading to abuse. Not least, massive data stores are more likely to be hacked into, exposing users' data to the public or to malicious parties.

It is important for the user to see a smart city system not only as being effective and reliable, but also as protecting the user's personal information [PER 15]. We mean here information about the user's location, current, past or scheduled activity, the user's habits and the user's social interaction. The system can protect the user's data by encrypting it with the user's public key (therefore only the user is able to decrypt it, using his/her private key), as well as by storing the user's data only on devices that are under the control of the user. This can be done by creating a network of the user's devices (or other devices that the user trusts) and keeping the data within this network.

If the user needs to give some personal information (PI) to another entity, the user must be able to control that transfer in detail. To help with that control and support the concept of not storing users' PI on other devices, the information should only be *pushed* from the user's device to the other entity (i.e. not *pulled* from the user's device by various entities), and the other entity should aggregate all information as quickly as possible to store, in the long term, only anonymized and aggregated data.

9.3. Solutions for managing context information

In the previous section, we argued about the requirements of smart city infrastructures and some features that are central to our approach: context-awareness, robustness and privacy. In this section, we will detail the methods that can be used to fulfill these requirements.

9.3.1. *Related work and projects*

The architecture presented in this chapter is designed on the foundation offered by our experience with previous architectures, namely AmIciTy, tATAmI and DESIA.

AmIciTy (Ambient Intelligence for the Collaborative Integration of Tasks) is a simulation platform for multi-agent systems formed of a large number of agents. The agents have a location in space and can only communicate with their spatial neighbors. The agents exchange pieces of information, characterized by several context-aware measures, indicating area of interest and importance of the information. The goal of the experiments was to observe how a piece of information that is inserted in the system reaches all agents to which it is relevant [OLA 11]. A secondary goal of the experiments was to show that the implemented agent behavior (using mechanisms of self-organization) leads to a scalable and robust system. Experiments show that information reaches a large number of agents (even agents not interested in it who will discard it, but not before disseminating it further). A piece of information may reach an interested agent by multiple and redundant paths. Local behavior means the system is scalable and the load of an individual agent does not depend on the size of the system.

tATAmI (towards Agent Technologies for Ambient Intelligence) is a platform for the development and deployment of multi-agent systems, focused on AmI applications. It enables a MAS developer to quickly specify the behaviors of an agent, by means of S-CLAIM – an AOP language with all the needed primitives for communication, input and output, and knowledge management [BAL 12]. The distributed setup of the system is specified in only one file and at one click tATAmI starts all agents on the appropriate machines and begins the execution. In recent developments, tATAmI agents can communicate using a variety of platforms (e.g. Jade, WebSockets and Web Services), with no changes whatsoever in their source code.

DESIA (Devices, Environments and Social networks Integration Architecture) is an architecture for the foundation of AmI applications that relies on three main elements [MAC 15]: the Personal Layer, implemented in each of the user's personal devices, is responsible for providing the services needed to support the application clients running in such devices and for collecting personal context; the Ambient Layer, implemented in each organizational physical environment, is responsible for providing functionalities to support and integrate the hardware and software entities in such spaces; the Cloud Layer, the core of the proposed architecture, is responsible for integrating data from large group of users and sets of organizations, gathering social web information and providing high level inference services. In each layer, there are several agents managing the layer's functionality, using information from a data repository in the respective layer.

The architecture presented in this chapter relies on the concepts about data dissemination that we have developed in the AmIciTy experiments, uses the local architecture developed with the tATAmI platform, and follows the requirements and philosophy of DESIA.

CONSERT [SOR 15] is a proposed middleware that resembles the function of our proposition – to facilitate the delivery of context information. However, it is not focused on a great level of distribution of the system.

In this research, we have also taken inspiration from existing products or projects, either as a model for the architecture or as an example of how we do not want things to work in a smart city.

At present, there are many existing services and applications that allow users to store and manage their data. Most of these applications rely on storing the user's data in *the cloud*, which has an opaque structure in which the user is not able to control how the data is used or moved within the cloud; an unknown location which can be in the same country as the user or on another continent; and a non-deterministic caching policy (usually, everything is replicated on PCs and downloaded upon request on mobile devices, but discrimination is done solely based on the machine's operating system and not on the actual usage and capabilities of the device). Among such services Google Drive[2] and Dropbox[3] are the best known. Although simple,

2 Google Drive – https://www.google.com/drive/.

3 Dropbox – https://www.dropbox.com/.

such a solution for distributed storage is both not dependable – the user may not be connected to the Internet at the time and the file may not be cached – and privacy invasive, the files can be inspected by private companies as well as local or foreign government entities. Moreover, the storage is file-based and not association-based. Google also offers Now[4], a service to deliver some types of relevant information proactively, at the time when they are needed. However, the same caveats apply.

Other initiatives attempt to build a distributed network for storage, messaging and email (such as Owncloud[5], Freenet[6] or Diaspora[7]), however these approaches are not interoperable, nor sufficiently popular.

9.3.2. *A local solution for a global result*

Looking into the requirements presented in section 9.2, we can infer some of the features of the solution.

The scalability requirements, SCL-1, SCL-2 and SCL-3, imply that the system must have a great degree of distribution. Any part of the network should be able to work autonomously; therefore, it is not feasible to rely on centralized components. This is not only for scalability concerns. When the user requests an information and if the information is on a faraway server (possibly in a different country), delays or unavailable resources would cause a break in the user's flow (requirement OPT-1). The information needed by the user should always be stored close to him/her, on the device or on devices that are close-by in the network. This means that the *system should be completely distributed* and the *information should be local.*

As we desire that the system is proactive and anticipates the user's needs (requirement CTX-4), it means that the potentially needed information must be kept close to the user at all times. However, needed information is related to the user's context and activity. Therefore, the system must store locally the *information that is local to the user's context.*

For the system to be able to store the potentially needed information close to the user, it must be able to use for storage the device(s) currently in use by the user. Conversely, the user's data should be kept only on the user's devices, to fulfill requirement PRV-1. The same devices must be used for a certain level of matching, to determine data about the current activity and to detect if information matches that activity or state. Therefore, what must be achieved is a system with a *light footprint,*

4 Google Now – https://www.google.com/landing/now/.

5 ownCloud – https://owncloud.org/.

6 The Freenet Project – https://freenetproject.org/.

7 The diaspora* Project – https://diasporafoundation.org/.

both in terms of memory and computation (also according to requirement CTX-3). Although offloading computation to other, more powerful, machines may be possible, the requirements state that processing and storage should be doable locally as well.

The primary tool that we have chosen to implement the proposed architecture is a multi-agent system (MAS) architecture. It can be used to implement interaction between autonomous entities, support self-organization mechanisms that offer robust behavior and have been researched as the method of choice to implement distributed artificial intelligence. By being naturally distributed and containing autonomous units, multi-agent systems are the obvious choice for our architecture.

Autonomy, the central feature of software agents serves the requirements that we have stated well, as it allows the perspective of the agent to be local. Autonomy means that, if need be, the agent must be able to work on its own, with the knowledge that it has. Similarly, we desire a system in which individual devices are able to hold the potentially interesting data if no other information is available on the network at the time.

The system should be lightweight on resource-constrained devices, but should be able to use all the power available on more capable machines. Then, it is important to have a great degree of flexibility, both in the agent behavior and in the storage of information, so that the capabilities of the devices are used optimally.

One adequate representation of knowledge and situation is by means of structures that preserve relations between concepts, such as graphs and graph patterns [OLA 15]. Such a representation should be common to all applications in the system (while remaining generic), fulfilling requirement CTX-2. A graph-based representation is also easy to understand by machines and humans alike (requirement CTX-3). Matching between patterns and graphs can be used to detect the current situation or interest toward pieces of information. Such a process is visually intuitive, helping with requirement OPT-2.

9.4. MAS-based application-independent middleware

Smart cities are extremely complex systems formed of potentially billions of components that are each autonomous, communicating, gathering and consuming information. We could group the functionalities that are implemented in the various devices in a smart city into several layers (as inspired by Seghrouchni [ELF 08]).

The system is based on the *hardware* layer, composed of all the devices that are part of the electronic environment: sensors, actuators, intelligent appliances, smartphones and tablets, workstations and servers. All devices are interconnected by means of a ubiquitous, secure *network* that uses various protocols and mostly wireless transmission. The data that are transferred are offered to the layers above in a uniform format and by means of standard protocols, assured by the *interoperability* layer. The context-aware, intelligent transfer of information, as well as specific context-aware services are offered by the *smart application* layer. The smart application layer interacts with the user by means of the intelligent *interfaces*, that support multi-modal, natural ways of communication with the human users. We argued that in the past the smart application layer could be further divided into two sub-layers: a generic information exchange sub-layer and a sub-layer for application-specific processes [OLA 13]. A graphical view of these layers is shown in Figure 9.1.

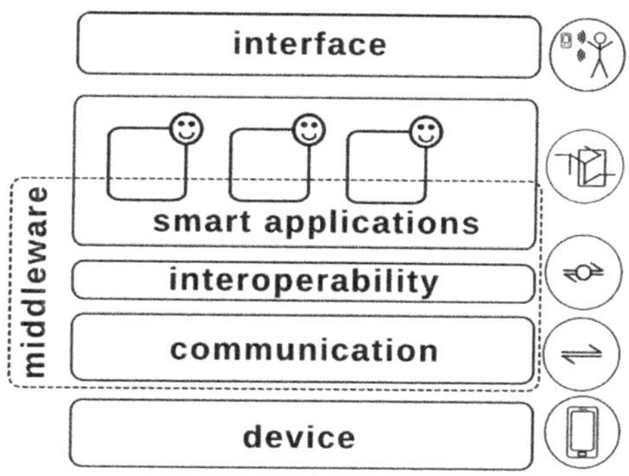

Figure 9.1. *A layered perspective on smart city applications, consisting of a hardware layer, a secure connectivity layer, a layer for interoperability of formats, a layer for smart applications and a layer for the interface*

Our aim is to build a middleware that will serve as an information exchange sub-layer, dealing with the context-aware, but generic (as in application-independent) transfer of information between the entities in the system. The middleware would be able to communicate with the user directly through the interface or would provide information to application-specific processes.

The middleware isolates several processes and features that are common to many smart applications: the delivery of context information to the interested users or applications; the detection of the user's situation and context; and in some cases, the decision on appropriate action that should be taken in the given situation. By isolating these processes, all there is left to do for users and applications is to "insert" information into the system and this information will reach other appropriate users and devices. The only requirement is that information inserted into the system respects the representation that is proposed. It is, of course, the application that performs domain-specific processing.

In the design of this architecture, there were two main priorities: first, keep a high degree of distribution and decentralization – this works in favor of the system's availability, robustness and performance, especially coupled with the self-organization mechanisms integrated in the agents; second, isolate inside the middleware all processes related to the transfer of information between devices this makes the system more modular and easy to be used together with different components.

It is not required that all devices in the smart city use the same middleware for context-aware communication. It is possible to create bridge agents or devices that transfer communication from other system to the proposed middleware.

9.4.1. *Architecture*

The architecture that we propose for a system-wide context-aware middleware relies on a multi-agent system in which autonomous agents handle context information according to the user's situation and activity.

Conceptually, the system is formed only of software agents that are able to communicate. Technically, some other software components may be necessary to enable the communication, depending on the means of communication that is used. For instance, if communication would be done through Jade – a popular MAS deployment environment then, a Jade container should be started on each of the machines.

Each software agent (see Figure 9.2) implements one or more well-defined functionality and serves one entity in the smart city (its *owner*). The owner of the agent may be a user or an entity such as a company, an association or a government institution. An agent is associated with one particular *element of context*, considering four types of context [HEN 06]: spatial, social, computational and activity. The agent contains in its knowledge-base information that belongs (or is available) to its owner and that is related to the context to which it is associated. This information is expected to be relevant to the owner in the present or in the short-term future.

Each agent executes on a machine (its host) that is related to its activity. The machine may belong to the owner of the agent or to another user/entity that authorizes the agent to execute there. Agents are mobile and, according to their functionality, may move from one machine to another.

Regarding *knowledge*, an agent must be able to represent knowledge as relationships between pieces of information, especially the relationship between a piece of information and related context. It must also have the means to detect if a piece of information that is received is relevant to the agent's functionality. Each element in this representation must be tagged with access rights – which users or entities are allowed to access that information. To this end, we recommend, for the proposed architecture, a representation based on Context Graphs, using Context Patterns to model agent interest [OLA 15].

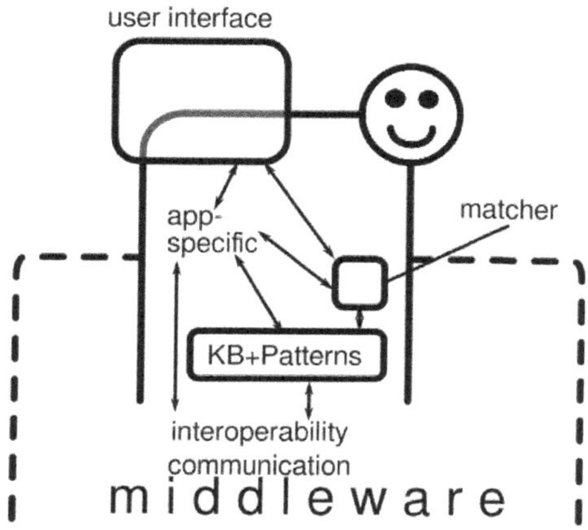

Figure 9.2. *A view inside an agent in the middleware. App-specific processes can perform queries directly on the middleware or are able to use the information (potentially using context matching) in the knowledge base. The matcher may be able to provide notifications and propose actions directly to the user, if the patterns allow it [OLA 15]*

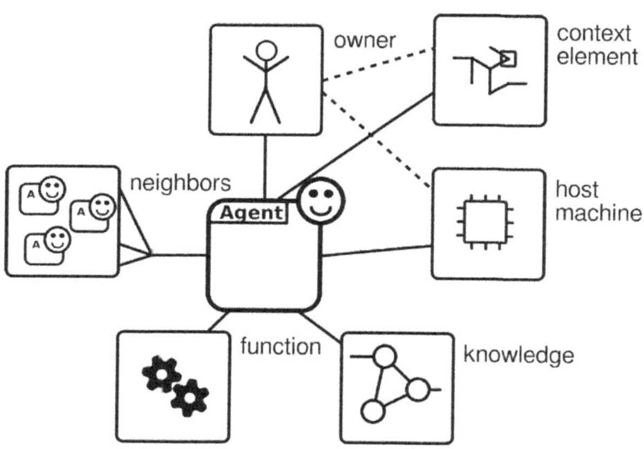

Figure 9.3. *A view of the elements connected to an agent in the middleware*

Some examples of agents that could be part of the proposed smart city agent-based architecture are presented in Tables 9.1 and 9.2 (see also Figure 9.3).

Function: an agent that provides the temperature reading from a room in a building	
Owner	The owner of the building and/or the entity renting the room or the space containing it
Associated context	The spatial context of the room
Agent knowledge	The current temperature in the room, potentially the recent history of temperatures in the room
Agent executes on	Smart device that features a microcontroller and a temperature sensor
Communicates with	Any agent that is authorized to receive the temperature
Function: an agent that gathers traffic information from drivers. This is one of a distributed network of such agents, which share the load and use partial global planning to coordinate	
Owner	The smart city municipality
Context	The spatial context of the area that the agent manages
Knowledge	Anonymous information about vehicle positions and velocity; aggregated information about driving patterns and density of vehicles on various streets
Aagent executes on	A powerful workstation in a local data center
Communicates with	Receives data from agents on the vehicles (if authorized by vehicle owners), collaborates with agents sharing the same overall task and reports results to the municipality
Function: an agent that manages the activities of a class of students their homework, their class activity, their results and their marks	
Owner	The school
Associated context	The group of students, as a class in the school, and the overall activity of the class
Agent knowledge	The information about the students, the schedule of the class
Agent executes on	A local machine in the room where the class is being currently held, or on a school server if the class has no current activities
Communicates with	The agents on the students' (with the permission of their parents) and teachers' devices and agents managing the activities of the school

Table 9.1. *(Series A) Examples of agents in a smart city, together with their main features – the owner, the associated context, the knowledge, the host machine and the agent's neighbors*

Function: an agent that manages a researcher's activity regarding a particular journal paper that s/he intends to write	
Owner	The researcher
Associated context	The activity of writing the journal paper
Agent knowledge	Information about the call for papers, topic, other papers related to the topic, experiments providing the results for the paper, collaborators on the paper and their expected contribution
Agent executes on	A device belonging to the user on which the user works or is likely to work on the paper (e.g. laptops or workstations, tablets (for reading), less likely the smartphone)
Communicates with	The agent scheduling the researcher's activity; in case of a paper written by multiple researchers, also the agents managing the activity of the other researchers, and an agent managing the whole activity of writing the paper
Function: an agent that assists a participant at a large social event (e.g. a rock concert)	
Owner	The organizer of the concert
Associated context	The concert (an activity)
Agent knowledge	Up-to-date information about the schedule, a map and navigation services for the facilities, indications to follow in case of emergency (e.g. fire)
Agent executes on	The participant's smartphone (if given permission; alternatively some of the information can be made available to an agent owned by the participant)
Communicates with	The personal agent of the participant; receives information from the organizer of the concert about scheduling updates and emergencies (e.g. the location of the fire)

Table 9.2. *(Series B) Examples of agents in a smart city, together with their main features – the owner, the associated context, the knowledge, the host machine and the agent's neighbors*

As we can see from the examples, each autonomous agent in the system has several features: the functionality, the owner, the associated context, the knowledge that it stores, the host machine and its topological neighbors (agents with which it communicates).

The functionality of an agent is described by two elements. As part of the middleware, the set of situation descriptions for situations are relevant to the agent. This can be done by using context patterns. These show what knowledge the agent is supposed to keep and provide to other agents, as well as what actions the agent should perform (for instance, in case some information is missing [OLA 15]). The second element is the application-specific functionality, which has access to the agent knowledge and is able to add higher-level knowledge to the knowledge base/graph.

The agent's knowledge is all the information that may be relevant to the agent's functionality. For instance, if the agent is managing an activity (it is associated with an element of activity context), it should store the information that is relevant to that activity. If the available storage or the bandwidth available for transfer are limited, the most relevant information must be stored.

The owner of the agent is fixed to the agent and is represented in the knowledge base accordingly. All the knowledge that the agent generates is marked as belonging to the owner and the owner is able to set permissions on it. It is safer to never be able to change the owner of an agent. If needed, the knowledge agent can be transferred to a newly created agent and the old agent destroyed.

The agent's associated context is an element of context that the agent is bound to and usually manages. It can be a (smart) space, a device, an activity (of a user, group of users or other type of entity) or the general information/overall activity of a user or entity. Associated context must not necessarily be related to the owner of the agent. Although the agent represents the interests of its owner, it may be associated with another user or activity (see the concert example).

The host machine on which the agent executes is related to its associated context. It executes on the machine that is closest to that context. For a user, it is the device that the user is currently using; for a smart space, it is a machine in that space, which has an access to the devices in the space; for a smart device, it is the device itself or a machine to which the device is connected; and for an activity, it is a machine in the smart space in which the activity takes place or a machine belonging to the user that performs that activity.

The neighbors of the agent are those agents that share a context with the agent. Because they share a context, it makes sense for the agents to exchange information related to that context. Thus, a topology of the agent system is formed, based on a common context. For example, the agent managing the sub-activities that are assigned to a user, as parts of a larger, group activity, should communicate with the personal agent of the user (shared social context the user), with agents of other users managing other sub-activities of the same activity (shared activity context) and with the agent managing the overall activity (also shared activity context).

In summary, the architecture relies on agents storing the information that will be needed and communicating relevant information to agents that share a context and have access to that particular data. This agent behavior has already been described at length [OLA 13].

9.4.2. *Generality of the design*

We argue that the presented architecture is general and application independent. Moreover, applications are able to adopt using the functionalities offered by the middleware to varying degrees. An application can just insert information into the middleware but not use information that arrives; an application can use the information but mark all generated information as private to the owner; or an application may use the entire functionality of information exchange.

The fact that the middleware is agent-based does not restrict the model of applications that use the middleware. Although an agent-based application may indeed use more features of the middleware, it can still be used as a unified database for the user's data, much like storage components are used by apps on mobile devices today.

Although it is arguable that the chosen types of context may not cover all situations, the architecture is not restricted to using only the specified types. Even so, the four types of context (activity, social, computational and spatial) cover a very wide range of situations. A fifth type of context would be the temporal context, but it does not make sense for an agent to manage a certain period of time, especially as there are agents that manage activities (which have a certain time span).

9.4.3. *Resilience in case of failures*

The design of the proposed architecture starts from the beginning to be robust and dependable even in the face of emergency or critical situations. However, there are also less critical situations that the architecture is able to handle various kinds of potentially distressful situations.

One of the main causes of distress when using applications is network unavailability or lack of quality. Most smart applications rely on the network to connect to the cloud and to centralized servers, both for the interpretation of commands or detection of context, and for retrieving the information, usually on request and not in an anticipative manner. Reliance on the network means that poor quality of the communication leads to user dissatisfaction, frustration and breakage of flow. Although some authors argue that the network is pervasive, even in the Western world, there are many areas with poor or no connection (in which the user may get when travelling by train or car) and roaming costs may hinder or prohibit network usage for many travelers.

A smart application would be able to overcome these obstacles by performing caching on the information, within the limits imposed by the capabilities of the device. If a user plans a trip in his/her calendar or moves toward an area (or means of transportation) that is likely to be served poorly by the network, potentially relevant information should be proactively cached. Moreover, if the user carries more than one personal device, local networking and load sharing means the user is still able to access one device from another even if there is no connection to remote servers and services.

As argued earlier, in such cases, distractions may occur which come from outside the user's context (e.g. the user finds a landmark and wants to find out more about it). In such cases, the shared context of users in the network (e.g. the spatial location) means that public interest information can be shared among their agents according to the rules of the proposed architecture. Message routing can be performed by an opportunistic network that doesn't require all user to be connected to remote servers.

9.5. Conclusion

A unified infrastructure for a smart city must at the same time offer information dissemination, be dependable and be robust, capable of offering protection to the users' personal information. We have defined a set of requirements for such an infrastructure and proposed an architecture in which applications rely on a uniform middleware for the management of user information. This middleware relies on keeping information local to the user's devices and managing information such that the needed information is available to the user at the right time.

There are many challenges that arise from this proposition and they will have to be solved in the future. Among them, a scarcity of simultaneously connected devices, and issues regarding information aggregation to offer anonymity.

Although the components of the middleware are implemented in various existing projects and frameworks (such as tATAmI or GraphMatcher), the middleware itself must be put together and tested. An additional challenge is introduced by the need to test the final result in a setup which is equivalent, in scale, with an entire city.

9.6. Bibliography

[BAL 12] BALJAK V., BENEA M.T., EL FALLAH SEGHROUCHNI A. *et al.*, "S-CLAIM: An Agent-based Programming Language for AmI, A Smart-Room Case Study", *Proceedings of ANT 2012, The 3rd International Conference on Ambient Systems, Networks and Technologies*, August 27–29, Niagara Falls, Ontario, Canada, vol. 10 of *Procedia Computer Science*, Elsevier, pp. 30–37, 2012.

[BRÉ 07] BRÉZILLON J., BRÉZILLON P., "Context Modeling: Context as a Dressing of a Focus", KOKINOV B., RICHARDSON D., ROTH-BERGHOFER T. *et al.* (eds), *Modeling and Using Context*, vol. 4635 of Lecture Notes in Computer Science, Springer Berlin Heidelberg, pp. 136–149, 2007.

[CAR 09] CARAGLIU A., DEL BO C., NIJKAMP P., Smart cities in Europe, Research Memoranda no. 48, Vrije Universiteit Amsterdam, 2009.

[CHA 07] CHACHULSKI S., JENNINGS M., KATTI S. *et al.*, "Trading structure for randomness in wireless opportunistic routing", *Proceedings of the 2007 Conference on Applications, Technologies, Architectures, and Protocols for Computer Communications*, vol. 37 of *ACM SIGCOMM Computer Communication Review*, ACM, pp. 169–180, 2007.

[CSI 92] CSIKSZENTMIHALYI M., CSIKSZENTMIHALYI I.S., *Optimal Experience: Psychological Studies of Flow in Consciousness*, Cambridge University Press, 1992.

[DEA 11] DEAKIN M., AL WAER H., "From intelligent to smart cities", *Intelligent Buildings International*, Taylor & Francis, vol. 3, no. 3, pp. 140–152, 2011.

[DEY 01] DEY A., "Understanding and using context", *Personal and Ubiquitous Computing*, Springer-Verlag, vol. 5, no. 1, pp. 4–7, 2001.

[DUC 01] DUCATEL K., BOGDANOWICZ M., SCAPOLO F. *et al.*, Scenarios for ambient intelligence in 2010, report, Office for Official Publications of the European Communities, February 2001.

[ELF 08] EL FALLAH SEGHROUCHNI A., "Intelligence ambiante, les défis scientifiques", *Colloque Intelligence Ambiante*, Forum Atena, December 2008.

[EVA 11] EVANS D., The internet of things: how the next evolution of the internet is changing everything, report, CISCO, 2011.

[GIF 07] GIFFINGER R., FERTNER C., KRAMAR H. *et al.*, Smart cities-Ranking of European medium-sized cities, report, Vienna University of Technology, 2007.

[HEN 06] HENRICKSEN K., INDULSKA J., "Developing context-aware pervasive computing applications: Models and approach", *Pervasive and Mobile Computing*, Elsevier, vol. 2, no. 1, pp. 37–64, 2006.

[MAC 15] MACIEL C., DE SOUZA P.C., VITERBO J. *et al.*, "A Multi-agent Architecture to Support Ubiquitous Applications in Smart Environments", *Agent Technology for Intelligent Mobile Services and Smart Societies*, Springer, pp. 106–116, 2015.

[ELF 12] EL FALLAH SEGHROUCHNI A., OLARU A., NGUYEN T.T.N. *et al.*, "Ao Dai: Agent Oriented Design for Ambient Intelligence", DESAI N., LIU A., WINIKOFF M. (eds), *Principles and Practice of Multi-Agent Systems, 13th International Conference, PRIMA 2010*, Kolkata, India, November 12-15, 2010, Revised Selected Papers, vol. 7057 of Lecture Notes in Computer Science, Springer Berlin/Heidelberg, pp. 259–269, 2012.

[OLA 11] OLARU A., GRATIE C., "Agent-Based, Context-Aware Information Sharing for Ambient Intelligence", *International Journal on Artificial Intelligence Tools*, World Scientific, vol. 20, no. 6, pp. 985–1000, December 2011.

[OLA 13] OLARU A., FLOREA A.M., EL FALLAH SEGHROUCHNI A., "A Context-Aware Multi-Agent System as a Middleware for Ambient Intelligence", *Mobile Networks and Applications*, Springer US, vol. 18, no. 3, pp. 429–443, June 2013.

[OLA 15] OLARU A., FLOREA A.M., "A Platform for Matching Context in Real Time", ONIEVA E., SANTOS I., OSABA E. (eds), *Proceedings of HAIS 2015, The 10th International Conference on Hybrid Artificial Intelligent Systems*, June 22-24, Bilbao, Spain, vol. 9121 of Lecture Notes in Computer Science, Springer International Publishing, pp. 99–110, June 2015.

[PER 15] PERERA C., RANJAN R., WANG L. *et al.*, "Privacy of Big Data in the Internet of Things Era", *IEEE IT Professional Magazine*, 2015.

[RUL 15] RULE J.B., Commentary: 'Free' Internet services do have a cost: Your personal information, Portland Press Herald, Online (http://www.pressherald.com/2015/08/05/commentary-free-internet-services-do-have-a-cost-your-personal-information/), August 2015, Retrieved October 2015.

[SIN 14] SINGH S., Smart Cities – A $1.5 Trillion Market Opportunity, Forbes, Online (http://www.forbes.com/sites/sarwantsingh/2014/06/19/smart-cities-a-1-5-trillion-market-opportunity/), June 2014, Retreived June 2015.

[SOR 15] SORICI A., PICARD G., BOISSIER O. *et al.*, "Multi-Agent Based Flexible Deployment of Context Management in Ambient Intelligence Applications", *Advances in Practical Applications of Agents, Multi-Agent Systems, and Sustainability: The PAAMS Collection*, Springer, pp. 225–239, 2015.

[TEN 13] TENE O., POLONETSKY J., "Big data for all: Privacy and user control in the age of analytics", *Northwestern Journal of Technology and Intellectual Property*, HeinOnline, vol. 11, pp. 239–273, 2013.

[WEI 95] WEISER M., "The computer for the 21st century", *Scientific American*, Munn & Co., vol. 272, no. 3, pp. 78–89, 1995.

10

City of Santander

This chapter shows the current status of the city of Santander in the transformation process to become a global smart city, including the ClouT contributions.

10.1. Introduction

Nowadays, more than 50 percent of humanity lives in urban centers, and the majority of the world's projected population growth over the next 30 years will take place in cities, not only in rich cities but also poor ones. This high population density together with the status of their infrastructures explains the high impact of cities on the environment: consuming two-thirds of the world's energy and being responsible for the majority of the harmful greenhouse gases. In addition, taking into account the economic crisis, urban areas management has become one of the most important development challenges of the current century [MIT 14].

Most cities are focusing their efforts on improving issues such as the management of public services, including street cleaning, waste management, green areas maintenance and street lighting; citizen safety and transportation. This common city context makes local authorities think about new urban management models.

As the center of the social, administrative and economic activity, cities shall focus their efforts on ensuring a more efficient urban resource management, improving their capacities of attracting and retaining not only talent but also financial resources.

Chapter written by Sonia SOTERO MUÑIZ and José Antonio TEIXEIRA VITIENES.

Santander City Council has been characterized by its eagerness to provide a more efficient city management closer to the citizens through the use of Information and Communication Technologies.

Santander is in the vanguard of innovation when talking about smart cities. According to the IDC [IDC 12] report, Santander is considered one of the two most intelligent cities in Spain, because of a very strong performance in the smartness dimensions, specifically for smart buildings and smart energy and environment. It has provided a position of leadership in the smart cities paradigm, creating a competitive advantage, thanks to a combination of research, innovation, knowledge and technology. The city council is trying to consolidate this position, being convinced that high levels of innovation will generate positive results for businesses and contribute to the growth of our economy.

The city has been proactive in trying to position itself as a future smart city through its SmartSantander (see Chapter 8) initiative. Though there is a lot of work going on around the world in the smart cities paradigm, there is no location with a facility like that used in Santander.

Making city's systems and services smarter saves cost and increases efficiencies, contributing to a more sustainable city, while positioning us for a long-term economic growth. Innovation in management and governance provides qualitative improvements to the public services.

In this scenario, the adoption of ICT technologies allows us to improve how our city works and stimulates a thriving, knowledge-driven economy. Santander City Council is improving the current service delivery capabilities by making their services (mobility, transport, street and public lighting, cultural and touristic information, e-government, waste management and irrigation) smarter. Santander will become a better place to stay for citizens and visitors, activating its economic development and reinforcing the international projection, because of several initiatives carried out within the city.

The development of Santander as a smart city will take several years to be completed. During this time, we expect to attract a large number of private companies. They will undoubtedly bring talented people to our city, but we will also hire highly trained people, thanks to the joint collaboration with our university and other companies based in our area.

Figure 10.1. *Outdoor parking and environmental monitoring used architecture. For a color version of the figure, see www.iste.co.uk/elfallah/smartcities.zip*

Retaining talented young people in Santander and growing the city's economy is one of the main goals of Santander City Council. As the majority of our initiatives are carried out in the area, the development of new services and applications as well as the deployment of the infrastructures is having a great impact that bodes well for new jobs and other career opportunities.

Santander City Council has developed its strategic and innovation plan 2020 – a document drawn up with the participation of stakeholders within the city of Santander. It covers the main principles and general governing objectives for the period 2012–2020, aiming at situating Santander in the vanguard of science and innovation, promoting a new business model based on an effective public–private partnership collaboration that integrates technological thriving industry and generates an environment that is favorable to the research and development investment.

Santander has shown a high level of commitment to the research, development and innovation, contributing to a more sustainable productive model as well the economic development of the city in the near future. Furthermore, to create and facilitate this business model, the municipality will support the generation of new ideas, commercialization, entrepreneurship and business formation within our city.

It is important to mention that Santander has stimulated from the beginning of the creation of the "Red de Ciudades Inteligentes en España, RECI" (Network of Smart Cities in Spain) [REC 12]. This network aims at creating a legal, political and regulatory framework to support local governments and authorities in the process of the transformation toward moderns smart cities. This framework will allow city members to share their experiences, carried out within the different smart initiatives, taking benefit from other projects, saving costs, enriching municipality services and establishing new task forces to obtain diverse ways to finance the different initiatives as well. Santander major is the president of this Network of Smart Cities in Spain. Besides Santander, other 64 cities are members of this initiative.

Local companies, committed to developing a thriving industry founded on technological innovation, demanded a regulatory framework for the wide variety of products and projects to be developed within the smart-city paradigm. Santander is part of the Technical Standardization Committee AEN/CTN 178 for smart cities, guaranteeing the standardization of a new urban development model that will permit sustainable and smart management of cities. It was launched due to the agreement signed by AENOR and the Secretary of State for Telecommunications and the Information Society of the Ministry of Industry, Tourism and Trade (SETSI) signed to promote standardization in the field of smart cities, and in this way contribute to their implementation and deployment in Spain.

Because of the commitment with the research and development, the Spanish Government recognized Santander as the "City of the Science and Innovation" in early 2011. For 3 years, the city will be part of the "Impulso Network", a network of cities of science and innovation that have received this distinction.

10.2. ClouT project

Santander city is one of the pilot cities of the ClouT project (see Chapter 5). The purpose of field trials planned for Santander is to start carrying out valuable services to the citizens and, at the same time, to collaborate in researching a sustainable model of a smart city sensing platform.

Two field trials are being used: the participatory sensing for city management and traffic mobility management.

The participatory sensing for city management adapts and extends functionalities of Santander participatory sensing applications to be interoperable with the ClouT platform, which will leverage this application to increase its scalability property by providing big data storage, and fast and secure access to event data, thanks to the Cloud computing features provided by the ClouT project.

Traffic mobility management enables citizens and visitors to get access to an enhanced urban mobility experience and to leverage city transportation resources efficiently. In this sense, a vast amount of information generated in the city associated with the use of public transportation, connected to user events and preferences and to other types of data to be considered such as environmental parameters, is processed accordingly to generate real time alerts and events about urban traffic in the city, which could be highlighted in a map with its associated location, as well as to show the user the best choices for accessing to the corresponding destination.

Information retrieved by IoT devices used in the city and municipal service information are accessible from the Santander Open Data Platform [SOD 14], including 73 data catalogs that are classified into six categories: transport, urban planning and infrastructures, environment, science & technology, tourism, culture and leisure.

The following sections include a more detailed description of both field trials.

10.2.1. *Participatory sensing for city management*

The purpose of this field trial is to improve the participatory sensing application, called the pace of the city, which is currently offered by the city of Santander as a service for citizens and tourists, and works in the following ways:

– a user of this mobile application may generate events that occur within the city associated with a determined topic, such as cultural acts, malfunctioning city services, damage to city property, traffic incident and others. Additionally, each user may subscribe to different types of events to receive an alert when one of these is generated by other users;

– these events are received by the municipality, and, according to the topic they are associated with, they are sent to the corresponding department within the municipality;

– this department creates an internal incidence associated with the received event. Additionally, each step done by this department to fix the received event is reported and may be tracked not only by the citizen who generates it, but also by any user of the mobile app.

Figure 10.2 illustrates the event lifecycle.

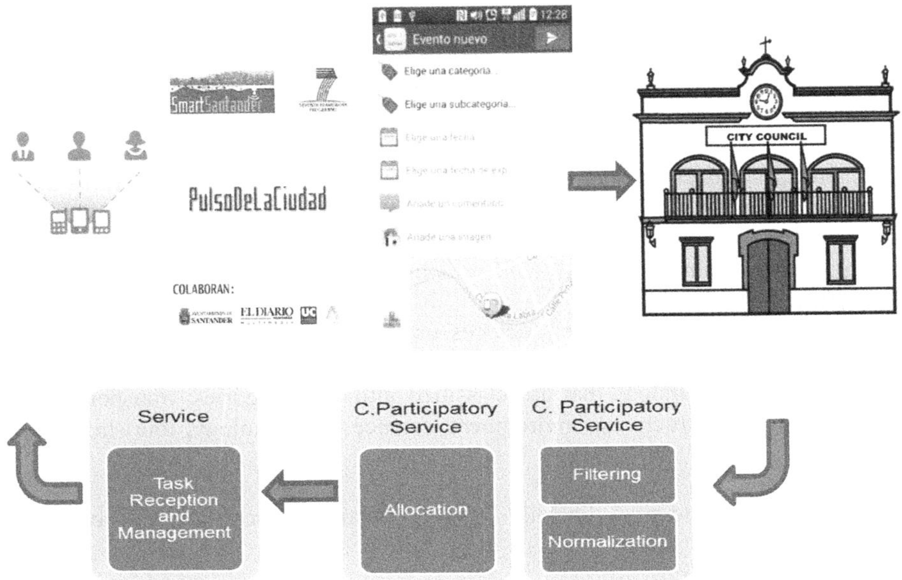

Figure 10.2. *Event lifecycle. For a color version of the figure, see www.iste.co.uk/elfallah/smartcities.zip*

In order to explain in a detailed manner the participatory citizens use cases, some scenarios exemplifying real situations have been defined, where the event registration, notification and subscription take place:

– Fernando has finished his working day and is going back home by car. Unfortunately, there is a lot of traffic and after 10 min being stuck in traffic, he notices that there is a hole in the road; therefore, he decides to report it by using the pace of the city app. As a consequence, his colleague Caroline, who also uses this application, receives an alert on her smartphone indicating the traffic incidence in that area, so she decides to take an alternative route home. Additionally, the corresponding department in the municipality receives the traffic event, opening the corresponding internal incidence and assigning it to the associated working team which moves there to fix it. As soon as the hole is repaired, the traffic event is closed by the municipal working team, which may be checked not only by Fernando but also by all the users of the mobile app.

– Sophie is a music fan, who is staying in the city for a vacation, and she wants to know if there are any musical events in the city during the afternoon. Thus, using her mobile phone, she discovers a cultural event: there will be a concert that night in the city's concert hall, also including the Website for buying tickets.

As can be derived from the aforementioned examples, the participatory citizen application presents a wide portfolio of both information producers (citizens, municipality and private companies), as well as information consumers (visitors, citizens, municipality and public services), thus involving a big percentage of the society.

Figure 10.3 (from the left to right and from top to bottom) shows different screenshots of the pace of the city application:

– Map: A map with all the events reported by users, including its geolocation. Each icon represents different types of events: traffic, beaches, street cleaning, trash cans, transportation, culture, weather, sports, tourism, water, parks and gardens and others.

– New event: How to report the event by completing the following information: topic, origin and expiration dates, comments, picture and geoposition.

– Events: Checking the detailed information associated with each reported event is possible.

– Incidence status: Event tracking is also available, making it possible to know the status of the event. Once the event has been solved, its status is finished, as can be seen in the last picture.

All the events reported by the pace of the city app appear at three different maps located at the SmartSantander Website (http://maps.smartsantander.eu/#page3), the mobile app and the Website version (www.elpulsodelaciudad.com), which belongs to a local newspaper, called El Diario Montañes. As can be seen, all the reported events are visible to all citizens and visitors. Additionally, actions done by the municipality to solve reported incidences are also visible. Therefore, it is important to highlight this visibility to focus main efforts of municipal services to ensure an efficient event management. Finally, social and cultural events are reported by users and also, the local newspaper which updates this category daily, to inform people within the city about concerts, exhibitions, etc.

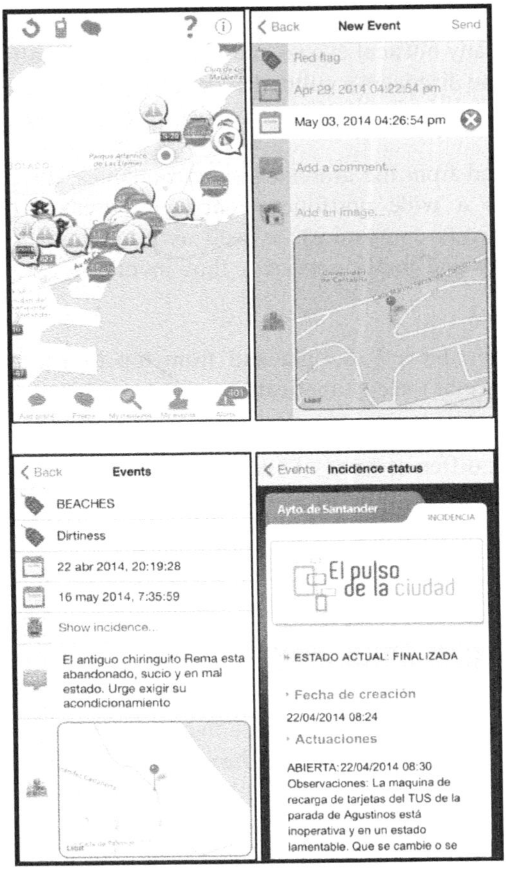

Figure 10.3. *Tracking of reported events. For a color version of the figure, see www.iste.co.uk/elfallah/smartcities.zip*

In addition to the city event generation, this application also allows us to access the values measured by the sensors with which mobile phones are provided, thus offering the capacity of sending this information to the central server.

It is important to highlight that, although the generated events and the map shown in the previous figure are associated with Santander, this functionality may be extended within the participatory sensing application, mainly attending to the following:

– Integration and usability: the associated information would be also stored in the Cloud solution for both sensor measurements information, as well as the event-related information (i.e. for storing pictures associated with events).

– Scalability: the solution may be extended from Santander to the remaining cities which participate in ClouT project (Genova, Mitaka and Fujisawa), thus associating each event with the corresponding municipality according to the GPS position where the event takes place. In this sense, the incidences that are reported in any of these four cities would be addressed, managed and solved by its corresponding municipal departments.

– Functionality: the user would be able to subscribe to the events that occur in a determined zone determined by the GPS position, for example, the user could subscribe to the events associated with the city in which he or she is. Additionally, it could be possible to define more specific topics depending on the particularities of each city.

In addition to technical issues, and considering that this application implies the participation of different and varied stakeholders, it is needed to address its social and economic impact.

From a social point of view, the creation of a mobile application allows the involvement of citizens within the smart-city concept, thus making them participants in, and contributors to, its development and improvement. Additionally, tourists and visitors will benefit from current application that contains information referring to the city and which is interactive in terms of addition and consumption of information. In this sense, the municipality is also completely involved in the development and use of this application, thus working hard to decrease the response times as well as improving the quality of the offered services. Finally, companies that offer public services, as well as other private companies, for example local newspapers, also state as an important part, thus updating the incidence status or including news related to this application.

From an economical point of view, this participative collaboration among all the stakeholders may be translated in the decrease in the expenses associated with the setting and reparation of the corresponding incidences, as the public services are directly subscribed to the incidences associated with the topics related to them.

10.2.2. *Traffic mobility management*

As can be seen in Figure 10.4, different information sources are feeding the processing unit, thus offering event monitoring and route estimation in terms of providing users with all the useful data related with the origin and destination points.

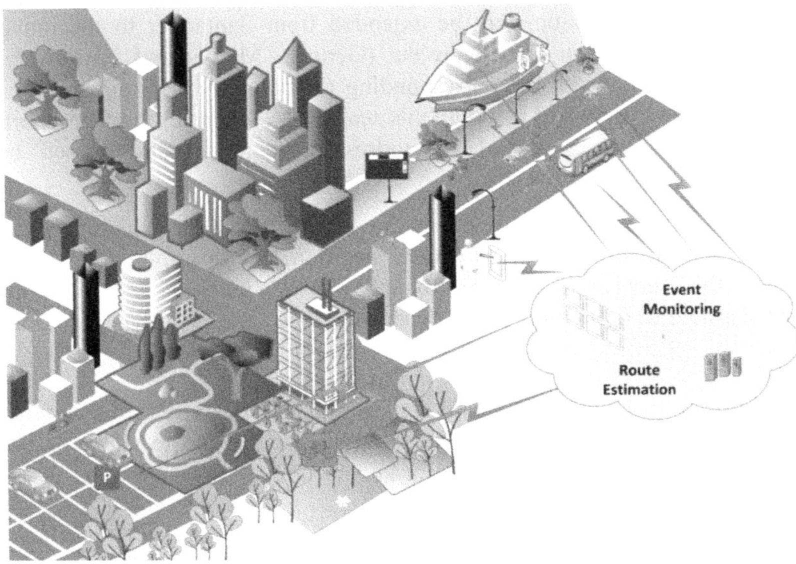

Figure 10.4. *Traffic mobility management field trial*

To offer added value services to their citizens, cities should provide the capacity to gather and process all the information generated by legacy and new data sources within each city context. In this sense, for this specific field trial, the main need to be addressed is the improvement of mobility within the city of Santander, not just considering information associated with traffic and transport, but going a step beyond and using the information generated by environmental devices located at fixed locations and mobile vehicles, as well as those events generated by users, thus offering an enriched solution to the users and fulfilling needs of the city in terms of providing more efficient mobility solutions.

This field trial is intended to gather and process information from different data sources related to transport and mobility within the city of Santander. In this sense, information about public bus fleets, such as their schedule, position, speed, estimated arrival time, crowdedness, expected travel time, as well as information related to other means of transport such bikes, taxis and trains, will be leveraged and combined on a single application. All these data are also combined with information regarding traffic parameters, such as speed, degree of occupancy and indoor and outdoor parking lots, as well as with environmental indicators such as CO_2, NO_2, O_3, noise and temperature; all of them provided by the static devices at main areas of the city or on mobile devices such as public vehicle or the citizen's smart phones.

Additionally, users' preferences and events generated by them will also be handled by this application, as well as the information directly gathered by the municipality that will be offered by the Santander Open Data Platform [SOD 14].

In order to show in a clear way the main object of this field trial, the following examples of use can be considered:

> *On a lovely sunny day, Pedro wants to spend the day at the beach with some friends. Therefore, he decides to use the traffic mobility management application to check which is the best way to get to Sardinero beach. The app retrieves the following information:*
>
> *– On the way to the beach, passing the house of his friends, there are some road works; as a consequence, traffic is slow on this road.*
>
> *– In the bike parking-lot close to his home, two bicycles are available. However, due to road works in this street, there are high levels of noise; so, he would prefer to avoid this route.*
>
> *– Looking at the free outdoor/indoor parking lots located near the beach, it seems that it will be complicated to park as the rate of parking spaces becoming available is quite low.*
>
> *– The next bus arriving at the closest bus stop is in 10 min and the following one will be in 40 min.*
>
> *Considering the previous information, Pedro finally decides to take the bus as it is the most comfortable option considering the road works and the difficulty of parking the car.*

As it can be derived from the previous example, this application would provide all the current information in the vicinity, as well as the possible routes, making it easier to make a decision about the best choice to get to the destination from the starting point.

As can be seen in Figure 10.5, different information about availability in the selected area of bus stops and the next buses to arrive, taxis available in the closest taxi ranks and bike availability in the public bike parking-lots in the vicinity has been displayed. This figure is just an example as information on available parking spaces at the destination, traffic density in the possible routes between source and destination possible road works in some of these routes could also be mentioned.

Figure 10.5. *Screenshot of the traffic mobility application. For a color version of the figure, see www.iste.co.uk/elfallah/smartcities.zip*

Finally, according to the scope of this field trial, the main stakeholders to be considered are indicated as the following:

– User (citizen): Citizens and visitors are provided with efficient routes according to traffic and environmental parameters, as well as to the suitability of the mean of transport to be used. For citizens, parameters like their scheduled routines and the routes of their friends can also be considered to calculate the most convenient route for a determined user. For visitors, preferences according to visiting more important points of interest of the city can also be taken into consideration to get efficient routes. It is clear that shortening and optimizing the transportation time has a clear impact on improving the city perception of users. Furthermore, users could monitor different events that occur in the city in a graphical way.

– Developer (third-party service provider): For third-party service developers, several applications can be developed by offering different alternative routes to the users according to the specific conditions indicated by them, such as the traffic circumstances, environmental care policies or the user preferences.

– System Administrators (Traffic Company/Municipality): Logically, the efficiency of transportation companies will be increased, as an application could be designed for balancing (in real time) the needs in terms of vehicles, staff, considering user requirements and preferences to user requirements (i.e. good weather implies more people going to the beach), specific city events (shows, sport events) as well as traffic/parking conditions, thus allowing a dynamic resource allocation and avoiding inefficient performance peaks and valleys. On the other hand, the municipality can offer a real time map of the status of transportation within the city, including colored alarms and events, thus allowing a dynamic reaction to the changing situations that can occurr throughout the day. Additionally, these applications could contribute to envisaging a future direction of city planning.

10.2.3. *Conclusions*

In the case of participatory sensing for city management, the Santander municipality provides a free application, the pace of the city, to citizens and visitors which allow them to generate events that occur within the city associated with different topics. Events associated with problems or malfunctioning municipal services are received by the Municipality, which will be in charge of fixing them and reporting the status of the event. Finally, these events are visible not only by the user who generated them, but also by all users of the application. Therefore, it is important to focus main efforts of municipal services to ensure an efficient event management.

The mobility application is able to gather all the different data sources that are providing information of different natures (environmental, traffic and parking), to process them accordingly and to show suitable information in a friendly way (through a website) to the user, according to his/her preferences. Additionally, to make application more efficient, new data sources will be included (i.e. filtered information from SNS), route selection according to the user preferences (i.e. assigning weights/priorities to corresponding measured parameters), as well as time estimation of arrival to the corresponding destination according to the route chosen by the user. Once application is more robust, it will be launched to be tested by some set of users and, finally, to be available for the citizens.

Finally, we will try to take advantage of the analyzed Cloud computing issues to improve service efficiency and performance of both participatory sensing for city management services as well as transport mobility management field trials, in terms of different features such as distributed processing, load balancing and data redundancy.

10.3. Bibliography

[IDC 12] INTERNATIONAL DATA CORPORATION REPORT, *Smart Cities Analysis in Spain 2012 – The Smart Journey*, 2012.

[MIT 14] MIT Technology Review, Cities Get Smarter, Business Report, 2014.

[REC 12] RED ESPAÑOLA DE CIUDADES INTELIGENTES, www.redcuidedesinteligentes.es, 2012.

[SOD 14] SANTANDER OPEN DATA PLATFORM, www.datos.santander.es, 2014.

11

Fujisawa, Towards a Sustainable Smart City

Fujisawa city and nearby areas have various sightseeing spots such as Enoshima Island, Shonan seacoast, Kamakura Buddha and others. Controlling tourists in the area is one of the important issues in Fujisawa city. During long holidays, a considerable number of people gather in specific areas in the city, which causes traffic jams or accidents. Providing information to move tourists into quieter areas is necessary. Another issue is to provide valuable information in emergency situations. Fujisawa is a seaside city, rendering it prone to earthquakes and tsunamis. Detecting such information and quickly providing information is needed. In terms of city resource management, location information of garbage trucks is important for Fujisawa municipality to plan effective routes and to collect garbage from the city, so as to reduce CO_2 emission. Providing air pollution sensor information measured by sensorized garbage trucks is also helpful for citizens' health. This chapter presents the challenges faced in changing Fujisawa city into a sustainable smart city.

11.1. Introduction

Fujisawa is located near the sea coast in Japan, so there is a high risk of earthquakes and tsunamis. In addition, Enoshima Island and Kamakura (city neighboring Fujisawa) are famous sightseeing places in Japan. Thus, it is necessary to provide real-time information to citizens as well as to tourists. For this context, we created an application called Enoshima Info Surfboard for the purpose of safety and emergency management use. In addition, Fujisawa city would like to invite more tourists to Enoshima Island from Kamakura. To motivate people to visit Enoshima Island, in addition to Enoshima Info Surfboard, we created another application called Smile Coupon for citizen health and to enhance pleasure of use.

Chapter written by Takuro YONEZAWA.

We used these two systems from October, 2014 and continue to operate them. The map around Enoshima and Kamakura is presented in Figure 11.1.

Figure 11.1. *Field trial area map*

On the basis of these systems, in the context of smart-city resource management, we designed and implemented a new mobile environmental sensor system installed in garbage trucks in Fujisawa city. We also created software which monitors the GPS location of each vehicle as well as several environmental sensor values such as pollen counts, PM2.5, and air contaminants. This information is visualized on a map to provide the operational status of garbage trucks and citywide atmospheric information to Fujisawa municipality and citizens.

11.1.1. *Sensorized garbage trucks*

To recognize whole city environmental information in a sustainable way, we designed a mobile atmospheric sensor system and installed it in garbage trucks in Fujisawa city. In Japan, many people are suffering from hayfever due to microscopic particles of pollen. In addition, the annual average concentration of PM2.5 is increasing year by year. Recently, many smart city projects have used many sensors in cities. However, this approach costs too much for deployment and management. Thus, we have to consider another approach for realizing a sustainable way of sensing. In this field trial, we focused on publically-owned vehicles, specifically garbage trucks, which are owned by the municipality. Garbage trucks run nearly every day in all cities. Thus, if we can use garbage trucks not only for collecting garbage but also for collecting various pieces of city information, it must be a sustainable way for city sensing. Monitored information can be used in various applications.

By having garbage trucks sensorized, we can provide citywide information about the atmosphere to citizens and visitors in Fujisawa city. The system has the following features:

– thirteen kinds of atmospheric sensors: CO, O_3, NO_2, air contaminants (two types of gases), dust, PM2.5, pollen, luminance, UV, temperature, humidity and ambient noise;

– the GPS location of a vehicle is measured every second and atmospheric sensor values are measured every 1 to 5 s depending on the sensor type. A series of measured sensor data are sent to a server every 30 s via 3G mobile networks;

– providing the position and operational status of each garbage truck as well as spatial distribution of each sensor value on a map in real time.

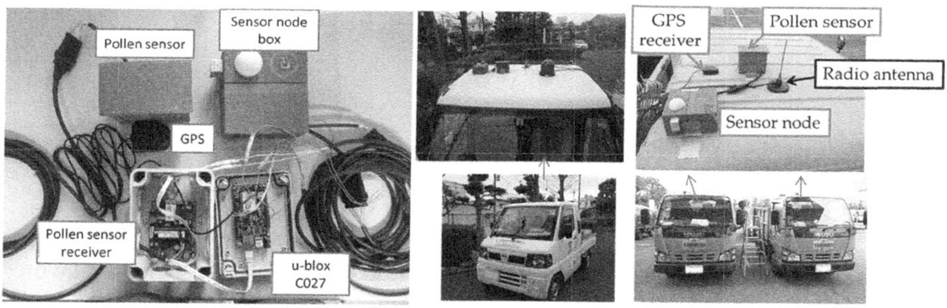

Figure 11.2. *Sensorized garbage trucks*

11.1.2. *Enoshima Info Surfboard*

Fujisawa city is one of the most dynamic cities in Japan. The population difference between daytime and nighttime is more than 100,000 people. In addition, there are many tourists in the seacoast area. Thus, sometimes there are heavy traffic jams both on the roads and railways. To offer a more efficient life for citizens and tourists, it is necessary to share and visualize city information including the dynamic Fujisawa context. Enoshima Info Surfboard can provide various items of city information from virtualized IoT in a cloud computing environment. We designed and implemented the system, and used it in Kamakura and Enoshima railway stations. Figure 11.3 shows the Enoshima Info Surfboard. It has railway following features:

– showing real-time city information such as weather, traffic, fish catches and so on from virtualized IoT in a cloud computing environment;

– integrated with digital signage system products;

– using projector light for ease of use;

– not only IoT information, but also the messages from the railway company can be shown.

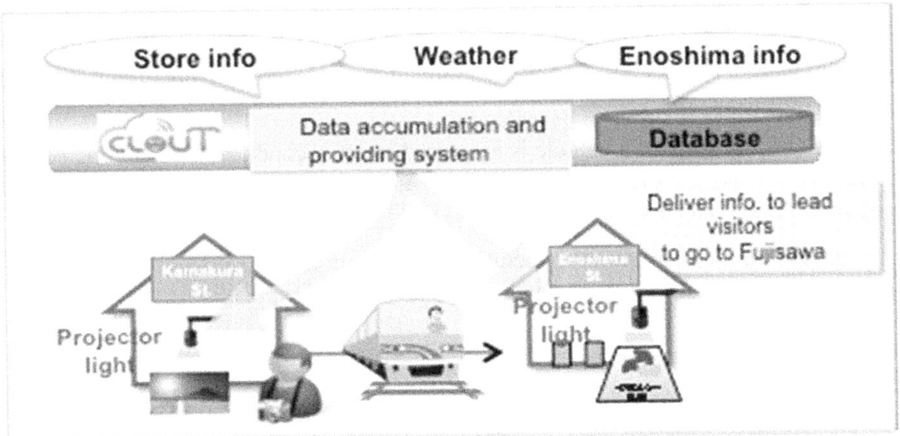

Figure 11.3. *Overall design of Enoshima Info Surfboard*

We also converted various websites as active sensors by using sensorizer technology. Not only sensor information, but also SNS data such as Twitter messages and messages from the railway company are managed as virtualized IoT. This enables us to integrate various pieces of information into Enoshima Info Surfboard very quickly. Therefore, we could easily reflect the demands of various stakeholders. For example, the Kamakura area is always full of people. Therefore, the railway company and Fujisawa municipality would like to transfer the tourists to the Enoshima area.

11.1.3. *Smile Coupon*

Smile Coupon was designed for tourists to change their behaviors. The Kamakura area always remains full of people. Therefore, the railway company and Fujisawa municipality would like to transfer the tourists to the Enoshima area. Smile Coupon is a dynamic coupon system that uses both local context and global context. As local context, we used the degree of people's smiles. According to smile, the discount level or exchangeable item of coupon is changed. In addition, the system also uses various IoT information such as weather and temperature for the coupon content. For example, in rainy days, we can set a higher discount level for the coupon. Figure 11.4 shows the overview of the Smile Coupon system. The system has the following features:

– using people's smiles as discount level of the coupon;

– easy and intuitive interface to take coupons. Just smile in front of the public display, and take a picture of it;

– integrating with various pieces of IoT information such as the weather. This enables the coupon system to be more dynamic according to the city context.

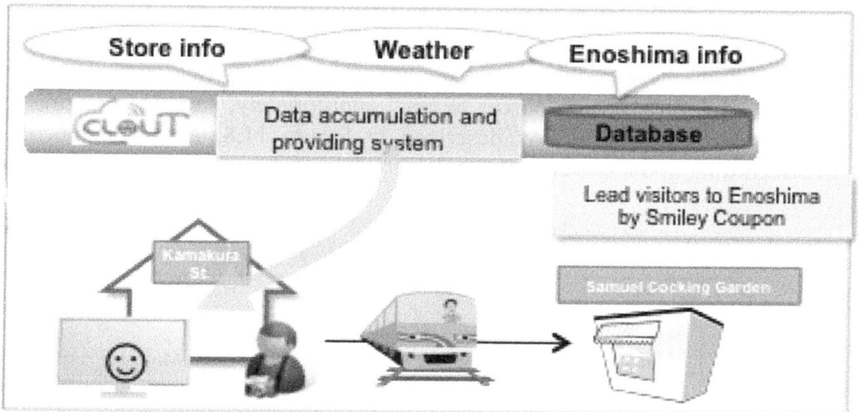

Figure 11.4. *Overall design of Smile Coupon*

We implemented the system and used it in Kamakura railway station. The coupon can be used in Enoshima Island. Therefore, it motivates people to move to Enoshima Island from Kamakura railway station.

11.2. Architecture and application domains

11.2.1. *Architecture with ClouT components*

Figure 11.5 shows integration on ClouT architecture (see Chapter 5) for the Fujisawa field trial. The system can efficiently use each functional component in CIaaS, CPaaS and CSaaS. In a CIaaS layer, the Fujisawa city data collector is based on sensorizer functionality (see Chapter 2) and the Data Transporter uses Uniform Access to IoT Data functionality in ClouT reference architecture. The sensor data from garbage trucks are also managed by the Fujisawa city data collector. In the CPaaS layer, we can use a sensor data healing mechanism and secure data storage functionality. In the CSaaS layer, the Event Recognizer can be based on City Event Finder functionality. On top of these components, Enokama Info Surfboard and Smile Coupon can be developed.

Details of the architecture are as follows. First, we register city sensor resources in the Resource Manager component. Then, according to discovered resources, the Fujisawa City Data Collector collects data for applications in the field trial. Collected data are then transferred to the upper layer of ClouT architecture, CPaaS, for using functionality of Dependable Data Access, Data Correct and Data Storage. Finally, in the CSaaS layer, data are analyzed for detecting city events and recognized events or data itself are presented via city resources such as public display. In addition, based on the detected city situation, the City Function Manager can also control several city resources to react appropriately, such as changing the sensing frequency of sensors. As described in Figure 11.8, each component is designed according to the ClouT architecture.

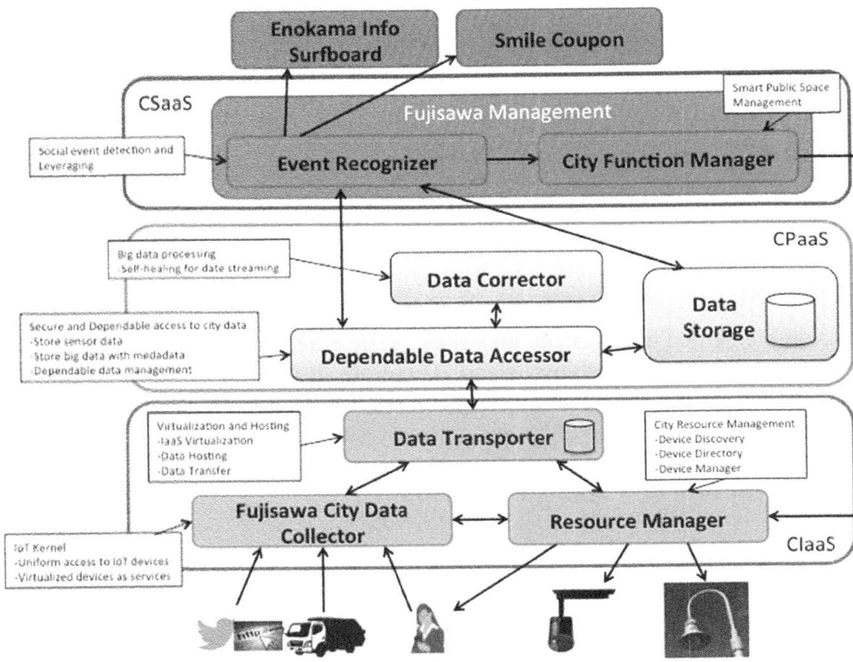

Figure 11.5. *Integrated architecture for the Fujisawa field trial*

11.2.2. *Components for implementation*

In addition to ClouT components such as sensorizer and event finder, we used open source software to implement Enokama Info Surfboard and Smile Coupon. Below is the list of reusable components:

– Smack – Java XMPP Software;

– Strophe.js – Javascript XMPP Software;

– Processing.js – Javascript animation library;

– OpenCV – image processing software;

– Clmtrackr – face detection software.

11.2.3. *Interaction among components*

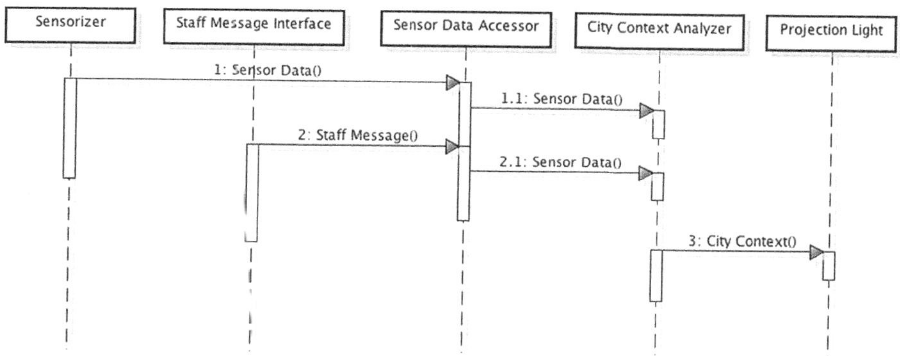

Figure 11.6. *Interaction among components for Enoshima Info Surfboard*

Figure 11.6 shows simple interaction among components for the Enokama Info Surfboard system. It has five main components – sensorizer, staff message interface, sensor data accessor, city context analyzer and projection light. Sensor data come from the sensorizer, staff messages and actual sensors are sent to the city context analyzer through the sensor data accessor. In addition, the recognized city context is used for information provision by using projection light.

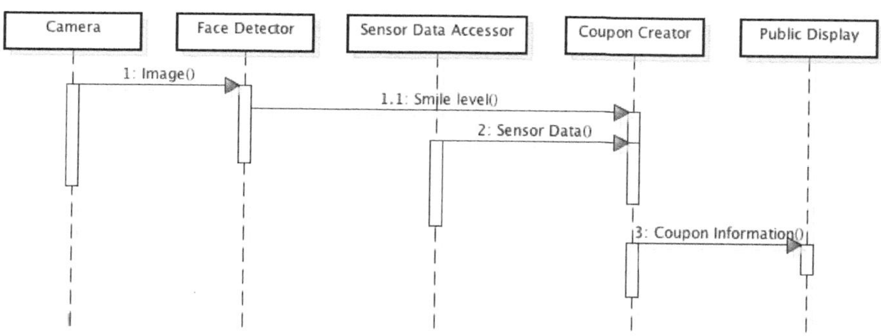

Figure 11.7. *Interaction among components for Smile Coupon*

Figure 11.7 also shows simple interaction among components for the Smile Coupon system. It has five main components – camera, face detector, sensor data accessor, coupon creator and public display. First, the user's face is captured by the public display's camera, and the captured image is analyzed by the face detector component. At the same time, the sensor data accessor provides various sensor data from the ClouT architecture, and the coupon creator decides the content of the coupon by using information of both the smile level and the sensor data such as weather information. Then, the public display shows the coupon to the user.

11.2.4. *Development scenario*

The Fujisawa field trial was started from the middle of October, 2014. We used the Enoshima Info Surfboard and Smile Coupon systems in Kamakura railway station. In terms of Enoshima Info Surfboard, we used Enokama Gateway – an information board in Kamakura. We attached a paper screen to the board, and we projected information on to the screen. As the projector, we used a Panasonic product called SpacePlayer. In the latter part of 2014, we also integrated our system with a digital signage system called NMStage. We developed API to enable the product to communicate with ClouT architecture. In terms of the Smile Coupon, we put a public display in Kamakura railway station.

As for the sensorized garbage trucks, we have refined the prototype system, developed in the first year and designed a new custom sensorboard in the middle of October, 2014. As a Windows tablet was used in a prototype system, we had to ask drivers to power up the tablet every time for data collection. In the second-year development, we simplified the overall system so that the sensor system automatically collects sensor data after the engine is started. We also resolved radio jamming from switching regulators on a sensorboard that interfered with radio communication between drivers and a station. We installed the new sensor system in three garbage trucks from the end of 2014 to the end of April, 2015 and also developed software which can monitor the location and operational status of each vehicle and visualize measured sensor data on the map.

The concrete scenario of the Fujisawa field trial is as follows:

1) Increasing the number of tourists from the Kamakura area to the Enoshima area: various pieces of information about Enoshima Island are projected. With the information, the tourists have an interest in the Enoshima area. In addition, we provide the Smile Coupon system, whose coupons can be used in the Enoshima area.

Therefore, we expect to increase the number of tourists in Enoshima Island from the Kamakura area.

2) Providing real-time safety information to citizens: in addition to physical sensor deployments, we used a number of websites as sensors. Some of the information included safety and security information such as earthquake or train accidents. This information can be sent to Enokama Info Surfboad in real time.

3) Providing interactive communication between citizens and railway company staff: not only sensor information, we also provide a message input interface to the railway company staff. It enables staff to provide various messages to citizens and enables communication between them.

4) Providing city resource and atmospheric information to the municipality and citizens: the Fujisawa municipality can monitor the location and operational status of each garbage truck. As the garbage trucks move along citywide routes, we can also provide detailed atmospheric information in the wide area of Fujisawa city along with information measured at existing stationary monitoring stations.

11.2.5. *Design and implementation*

In terms of a sensorized garbage truck, we designed a custom sensorboard and re-designed the entire system from the prototype developed in the first year. Figure 11.8 shows a block diagram of the system and Figure 11.9 shows a developed custom sensorboard. There is a sensor node for controlling several sensors on a sensorboard. We used a CIL virtual machine developed by NTTRD for implementing the program on the sensor node. As described in the previous section, we automated the system without requiring extra operations such as powering on a Windows tablet by drivers. For this purpose, u-blox C027 is adopted as a gateway for intermediating communication between a sensor node and a server so that the system starts collecting sensor data after the engine is started. The program running on u-blox/Linux gathers sensor data collected by the sensor node, pollen counts from the pollen sensor and GPS data from the receiver and sends a bunch of data to a remote server via 3G mobile networks every 30 s.

The remote server stores the sensor data from each sensorized garbage truck and can transmit the thinned-out sensor data to another cloud server via XMPP protocol. We have also developed software called Control Center which can monitor the current location and operational status of each vehicle and visualize sensor data on the map in real time.

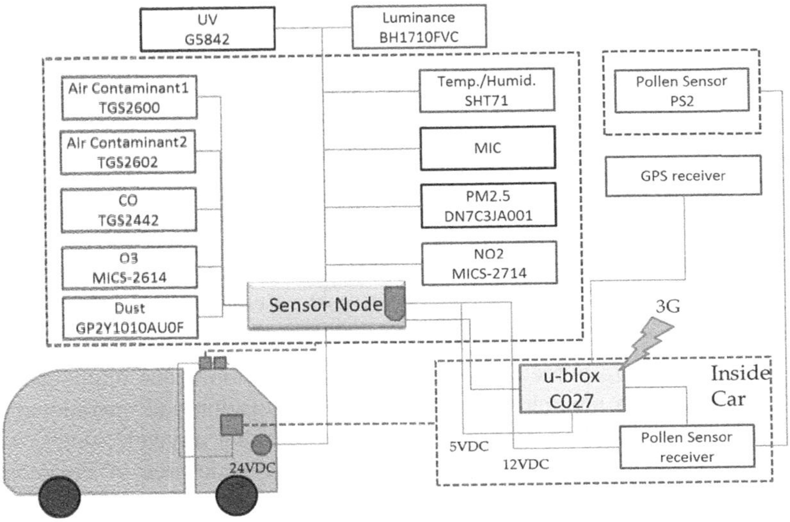

Figure 11.8. *Block diagram of sensorized garbage truck. For a color version of the figure, see www.iste.co.uk/elfallah/smartcities.zip*

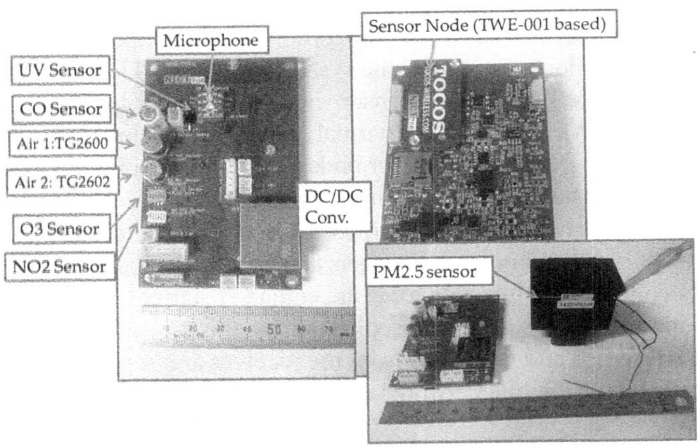

Figure 11.9. *Custom atmospheric sensorboard*

Figure 11.10 shows a screenshot of Control Center. Red circles indicate locations of each car, and the color of a trajectory indicates the value level of the selected sensor type. The higher (lower) the sensor data value, the color becomes more red (blue). It can also play back past retrieved data.

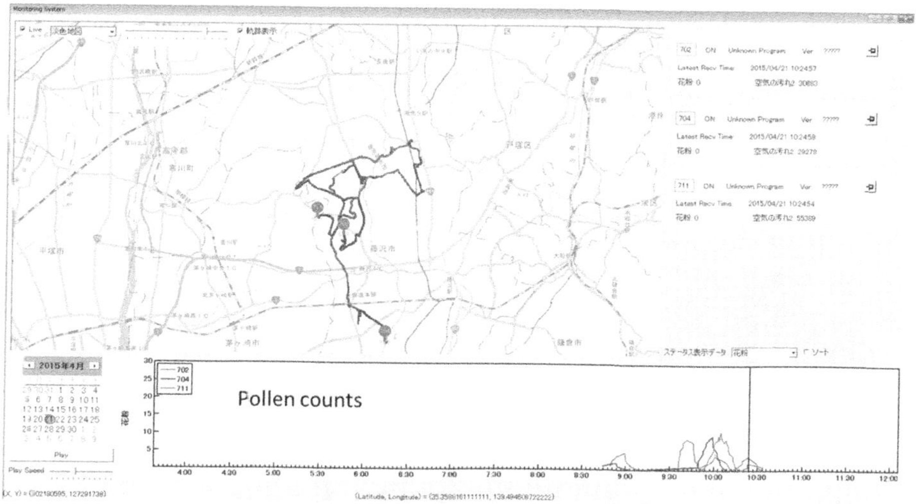

Figure 11.10. *Control center for monitoring car location and sensor data on a map. For a color version of the figure, see www.iste.co.uk/elfallah/smartcities.zip*

We designed and implemented Enokama Info Surfboard and Smile Coupon in Kamakura railway station. In addition, we also used the information board in the Enoshima railway station.

In terms of Enokama Info Surfboard, we mainly had two phases. First, we implemented the system without the existing product. Second, we implemented the system with the existing digital signage product (NMStage by Panasonic) by creating specific API for the product. NMStage provides two types of content. One is the static content such as event information, the other is the dynamic content received from the Sensor Virtualization Tool through a specific API. The contents, such as weather, notices about special events and information on Twitter, which are updated constantly, are displayed in real time.

Figure 11.11 shows an actual setup of Enokama Info Surfboard. We used three SpacePlayers (light projectors), so that we could provide several pieces of information at the same time. In Enoshima railway station, we also used two SpacePlayers and screens to provide information about Enoshima (see Figure 11.12). Existing works focused on the relationship between users and interactive public displays [MÜL 09b, MÜL 12, VOG 04]. However, these relationships will not be static, and will change according to the city nature and environmental information. Therefore, in this field trial, we focused on the city nature and environmental information and explored relationships between the user's activity and these parameters.

Figure 11.11. *Enokama Info Surfboard*

Figure 11.12. *Information board in Enoshima*

Table 11.1 shows the examples of content projected in Enokama Info Surfboard. We designed and implemented much content which is dynamically animated content according to IoT information, from ClouT architecture.

Projected Image	Content
	Weather information for Enoshima Island today. It includes maximum and minimum temperatures
	Weather information for Enoshima Island tomorrow. It includes maximum and minimum temperatures
	Current wind speed in Enoshima Island. A windmill is animated according to the strength of wind speed
	Information of sunset time in Enoshima. Enoshima is known as a beautiful sunset place. On sunny days, we can see Mount Fuji from Enoshima Island
	Information about ultraviolet radiation. Not only its strength, we also provide more information such as "damage to your skin is strong/weak"

<table>
<tr>
<td>
</td>
<td>Information of train operation such as delays and others. Kamakura station has three railway lines. It is important to provide information about trains to tourists and citizens so they can decide which train to ride</td>
</tr>
<tr>
<td>
</td>
<td>Information of tonight's star-gazing conditions such as "we expect good stars tonight" or "today we cannot see any stars in the sky"</td>
</tr>
<tr>
<td>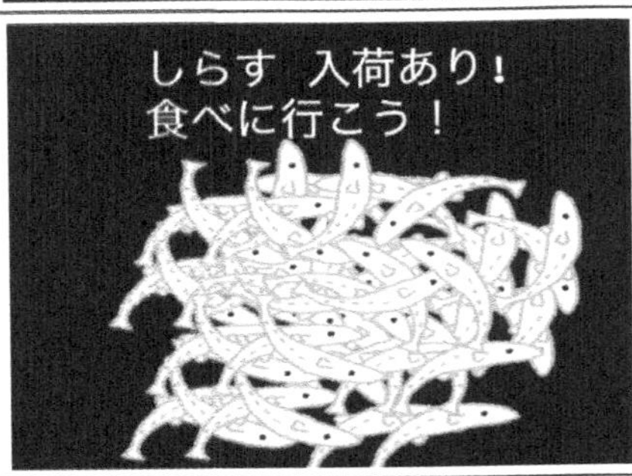
</td>
<td>Information about fish catches in Enoshima island. Many Japanese like to eat fresh fish. So, the information about fresh fish is fascinating for citizens/tourists</td>
</tr>
<tr>
<td>
</td>
<td>Information about "is today some kind of special day"? Each day has a special event such as a historical event. So, we provide information to citizens/tourists to use the information for daily conversation</td>
</tr>
<tr>
<td>
</td>
<td>Message from railway company staff. We provided an input system to the staff. The staff use the system to show various messages such as event information, casual talk and so on</td>
</tr>
</table>

Table 11.1. *Content list of Enokama Info Surfboard*

Figure 11.13 shows a Smile Coupon Setup. We used a 42-inch display attached to a web camera. In front of the display, we also placed instructions of how to use Smile Coupon. Figure 11.14 shows a screenshot of the Smile Coupon system. It shows the user's face, degree of smile and coupon content. According to smile level, we prepared several items for the coupon such as a towel, cup, figure and so on. Users can get a coupon just by taking a picture of the display. The coupon can be exchanged for the corresponding item at Enoshima Island. Figure 11.15 shows the place to exchange the coupon for the item. The place is located at the top of Enoshima Island. To get to the place, the users have to take a train to Enoshima station (20–25 min from Kamakura station) and have to walk from the station to the place in 30 min.

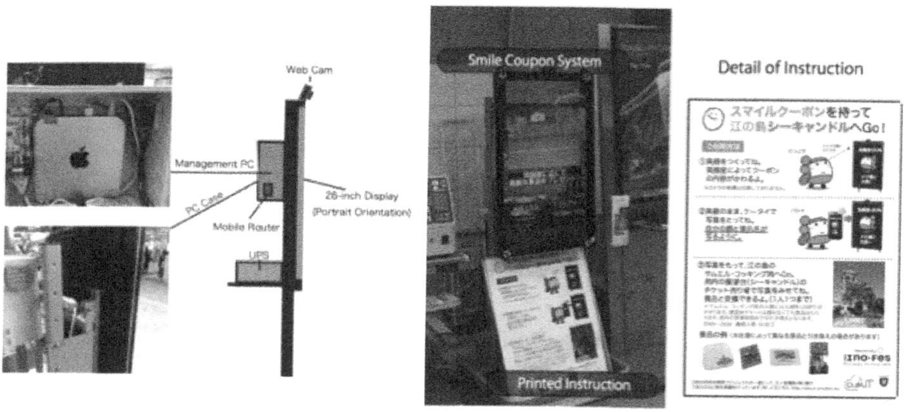

Figure 11.13. *Smile Coupon setup*

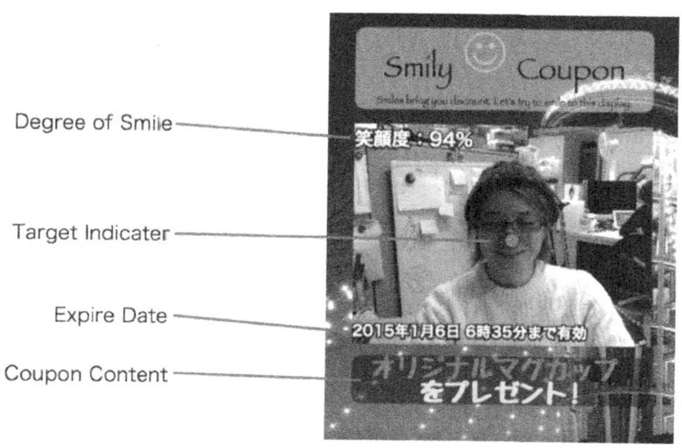

Figure 11.14. *Smile Coupon*

Figure 11.15. *Prize exchange booth*

11.3. Results

We reported the result of the Smile Coupon system operation in November and December, 2014. The number of actual coupons used (exchange number of coupon) was 64 (in November) and 33 (in December).

Item	November	December	Total
Mug Cup (Smile degree > 90%)	17	9	26
Mascot Figure (Smile degree > 80%)	20	17	37
Towel (Smile degree > 70%)	17	1	18
Post card (Smile degree > 60%)	3	0	3
Post-it (Smile degree > 50%)	4	2	6
Memo book (Smile degree > 40%)	3	4	7

Table 11.2. *Number of items exchanged*

Table 11.2 shows the number of items exchanged in each month. To exchange the coupon for the actual item, tourists have to ride a train and have to walk for 30 min. Since the travel is more than 1 h, we think this number of exchange items can be evaluated as our system working well. In terms of the coupon system in the public display research area, for example, Müller *et al.* used a coupon system for 1 year. They reported that the coupon was used only 37 times in their trial [MÜL 09a].

Compared with their system, our system seemed to be accepted by many people. We used a smile level and an actual picture of the face as the coupon. We think that this interaction enhanced the attachment of the coupon, and the people actually used it. We also conducted an interview with staff who worked to exchange coupons for items. They reported that most people seemed to enjoy our system. They also reported the age of people who used the coupon. The people were from various ages and groups such as couples and families. However, the number of older people was low. This is also because of the distance between Kamakura station and the exchange point. Based on this result, we are planning for the next field trial based on ClouT technology.

11.4. Conclusion

This chapter introduces field trials in Fujisawa city with ClouT architecture. To solve the problem of Fujisawa and also to enhance the life of citizens and tourists, we designed, implemented and used "Enoshima Info Surfboard" and "Smile Coupon" with various pieces of city information including "Sensorized Garbage Trucks". These systems are designed following ClouT architecture and implemented with several ClouT components. Therefore, it is easy to integrate with city resources, and also these applications can be transferred to other cities. As a result of field trials, we confirmed that our system can motivate some citizens to change their behaviors.

11.5. Bibliography

[MÜL 09a] MÜLLER J., WILMSMANN D., EXELER J. *et al.*, "Display Blindness: The Effect of Expectations on Attention towards Digital Signage, Pervasive Computing", *Lecture Notes in Computer Science*, Vol. 5538, Springer Berlin Heidelberg, pp. 1–8, 2009.

[MÜL 09b] MÜLLER J., KRÜGER A., "Mobidic: context adaptive digital signage with coupons", *3rd European Conference on Ambient Intelligence (Aml-09)*, Springer, pp. 24–33, November 2009.

[MÜL 12] MÜLLER J., WALTER R., BAILLY G. *et al.*, "Looking glass: a field study on noticing interactivity of a shop window", *Proceedings of CHI '12, ACM*, pp. 297–306, 2012.

[VOG 04] VOGEL D., BALAKRISHNAN R., "Interactive public ambient displays: transitioning from implicit to explicit, public to personal, interaction with multiple users", *Proceedings of UIST '04*, ACM, pp. 137–146, 2004.

List of Authors

Jander BOTELHO DO NASCIMENTO
French Commission for Atomic and Alternative Energies
France

Costin CAVAL
Pierre and Marie Curie University
Paris
France

Ahmed-Chawki CHAOUCHE
Pierre and Marie Curie University
Paris
France

Cédric DINONT
ISEN
Lille
France

Rémi DRUILHE
French Commission for Atomic and Alternative Energies
France

Amal EL FALLAH SEGHROUCHNI
Pierre and Marie Curie University
Paris
France

Adina Magda FLOREA
Polytechnic University of Bucharest Romania

José Antonio GALACHE
University of Cantabria
Spain

Etienne GANDRILLE
French Commission for Atomic and Alternative Energies
France

Levent GÜRGEN
French Commission for Atomic and Alternative Energies
France

Laurent HÉRAULT
VP Europe
CEA-Leti
France

Shinichi HONIDEN
National Institute of Informatics
Japan

Jean-Michel ILIÉ
Pierre and Marie Curie University
Paris
France

Fuyuki ISHIKAWA
National Institute of Informatics
Japan

Christophe MUNILLA
French Commission for Atomic and Alternative Energies
France

Luis MUÑOZ
University of Cantabria
Spain

Jin NAKAZAWA
Keio University
Japan

Andrei OLARU
Polytechnic University of Bucharest
Romania

Ferdinand PIETTE
ISEN
Lille
France

Djamel Eddine SAÏDOUNI
University of Constantine 2
Algeria

Juan Ramón SANTANA
University of Cantabria
Spain

Sonia SOTERO MUÑIZ
Santander City Council
Spain

Patrick TAILLIBERT
LIP6 and Thales Airborne Systems
France

Kenji TEI
National Institute of Informatics
Japan

José Antonio TEIXEIRA VITIENES
Santander City Council
Spain

Hideyuki TOKUDA
Keio University
Tokyo
Japan

Takuro YONEZAWA
Keio University
Tokyo
Japan

Index

T, V, W, X

Other titles from

ISTE

in

Information Systems, Web and Pervasive Computing

2016

BEN CHOUIKHA Mona
Organizational Design for Knowledge Management

BÉRANGER Jérôme
Big Data and Ethics: The Medical Datasphere

KITAJIMA Munéo
Memory and Action Selection in Human–Machine Interaction (Human–Machine Interaction Set – Volume 1)

LAGRAÑA Fernando
E-mail and Behavioral Changes: Uses and Misuses of Electronic Communications

MONINO Jean-Louis, SEDKAOUI Soraya
Big Data, Open Data and Data Development (Smart Innovation Set – Volume 3)

SALGUES Bruno
Health Industrialization

VENTRE Daniel
Information Warfare – 2nd edition

2015

ARDUIN Pierre-Emmanuel, GRUNDSTEIN Michel, ROSENTHAL-SABROUX Camille
Information and Knowledge System
(Advances in Information Systems Set – Volume 2)

BÉRANGER Jérôme
Medical Information Systems Ethics

BRONNER Gérald
Belief and Misbelief Asymmetry on the Internet

IAFRATE Fernando
From Big Data to Smart Data
(Advances in Information Systems Set – Volume 1)

KRICHEN Saoussen, BEN JOUIDA Sihem
Supply Chain Management and its Applications in Computer Science

NEGRE Elsa
Information and Recommender Systems
(Advances in Information Systems Set – Volume 4)

POMEROL Jean-Charles, EPELBOIN Yves, THOURY Claire
MOOCs

SALLES Maryse
Decision-Making and the Information System (Advances in Information Systems Set – Volume 3)

SAMARA Tarek
ERP and Information Systems: Integration or Disintegration
(Advances in Information Systems Set – Volume 5)

2014

DINET Jérôme
Information Retrieval in Digital Environments

HÉNO Raphaële, CHANDELIER Laure
3D Modeling of Buildings: Outstanding Sites

KEMBELLEC Gérald, CHARTRON Ghislaine, SALEH Imad
Recommender Systems

MATHIAN Hélène, SANDERS Lena
Spatio-temporal Approaches: Geographic Objects and Change Process

PLANTIN Jean-Christophe
Participatory Mapping

VENTRE Daniel
Chinese Cybersecurity and Defense

2013

BERNIK Igor
Cybercrime and Cyberwarfare

CAPET Philippe, DELAVALLADE Thomas
Information Evaluation

LEBRATY Jean-Fabrice, LOBRE-LEBRATY Katia
Crowdsourcing: One Step Beyond

SALLABERRY Christian
Geographical Information Retrieval in Textual Corpora

2012

BUCHER Bénédicte, LE BER Florence
Innovative Software Development in GIS

GAUSSIER Eric, YVON François
Textual Information Access

STOCKINGER Peter
Audiovisual Archives: Digital Text and Discourse Analysis

VENTRE Daniel
Cyber Conflict

2011

BANOS Arnaud, THÉVENIN Thomas
Geographical Information and Urban Transport Systems

DAUPHINÉ André
Fractal Geography

LEMBERGER Pirmin, MOREL Mederic
Managing Complexity of Information Systems

STOCKINGER Peter
Introduction to Audiovisual Archives

STOCKINGER Peter
Digital Audiovisual Archives

VENTRE Daniel
Cyberwar and Information Warfare

2010

BONNET Pierre
Enterprise Data Governance

BRUNET Roger
Sustainable Geography

CARREGA Pierre
Geographical Information and Climatology

CAUVIN Colette, ESCOBAR Francisco, SERRADJ Aziz
Thematic Cartography – 3-volume series
Thematic Cartography and Transformations – volume 1
Cartography and the Impact of the Quantitative Revolution – volume 2
New Approaches in Thematic Cartography – volume 3

LANGLOIS Patrice
Simulation of Complex Systems in GIS

MATHIS Philippe
Graphs and Networks – 2nd edition

THERIAULT Marius, DES ROSIERS François
Modeling Urban Dynamics

2009

BONNET Pierre, DETAVERNIER Jean-Michel, VAUQUIER Dominique
Sustainable IT Architecture: the Progressive Way of Overhauling Information Systems with SOA

PAPY Fabrice
Information Science

RIVARD François, ABOU HARB Georges, MERET Philippe
The Transverse Information System

ROCHE Stéphane, CARON Claude
Organizational Facets of GIS

2008

BRUGNOT Gérard
Spatial Management of Risks

FINKE Gerd
Operations Research and Networks

GUERMOND Yves
Modeling Process in Geography

KANEVSKI Michael
Advanced Mapping of Environmental Data

MANOUVRIER Bernard, LAURENT Ménard
Application Integration: EAI, B2B, BPM and SOA

PAPY Fabrice
Digital Libraries

2007

DOBESCH Hartwig, DUMOLARD Pierre, DYRAS Izabela
Spatial Interpolation for Climate Data

SANDERS Lena
Models in Spatial Analysis

2006

CLIQUET Gérard
Geomarketing

CORNIOU Jean-Pierre
Looking Back and Going Forward in IT

DEVILLERS Rodolphe, JEANSOULIN Robert
Fundamentals of Spatial Data Quality

Printed and bound by CPI Group (UK) Ltd, Croydon, CR0 4YY
16/09/2026
14963196-0001